ROUTLEDGE LIBRARY EDITIONS
BROADCASTING

Volume 9

BROADCAST TRANSMISSION ENGINEERING PRACTICE

BROADCAST TRANSMISSION ENGINEERING PRACTICE

WILLIAM WHARTON,
SHAUN METCALFE
AND
GEOFF C. PLATTS

Routledge
Taylor & Francis Group
LONDON AND NEW YORK

First published in 1992 by Focal Press

This edition first published in 2024
by Routledge
4 Park Square, Milton Park, Abingdon, Oxon OX14 4RN

and by Routledge
605 Third Avenue, New York, NY 10158

Routledge is an imprint of the Taylor & Francis Group, an informa business

British Library Cataloguing in Publication Data
A catalogue record for this book is available from the British Library

ISBN: 978-1-032-59391-3 (Set)
ISBN: 978-1-032-62248-4 (Volume 9) (hbk)
ISBN: 978-1-032-62254-5 (Volume 9) (pbk)
ISBN: 978-1-032-62252-1 (Volume 9) (ebk)

DOI: 10.4324/9781032622521

Publisher's Note
The publisher has gone to great lengths to ensure the quality of this reprint but points out that some imperfections in the original copies may be apparent.

Disclaimer
The publisher has made every effort to trace copyright holders and would welcome correspondence from those they have been unable to trace.

Broadcast Transmission Engineering Practice

William Wharton, CEng, FIEE
Shaun Metcalfe, CEng, FIEE
and
Geoff C. Platts, BSc, CEng, FIEE

FOCAL PRESS

Butterworth-Heinemann Ltd
Linacre House, Jordan Hill, Oxford OX2 8DP

PART OF REED INTERNATIONAL BOOKS

OXFORD LONDON BOSTON
MUNICH NEW DELHI SINGAPORE SYDNEY
TOKYO TORONTO WELLINGTON

First published 1992

British Library Cataloguing in Publication Data
Wharton, W.
Broadcast Transmission Engineering Practice
I. Title
621.382

ISBN 0 240 51335 5

Library of Congress Cataloguing in Publication Data
Wharton, William
Broadcast transmission engineering practice/William Wharton, Shaun Metcalfe, and Geoff C. Platts
p. cm.
Includes index.
ISBN 0 240 51335 5
1. Radio – Transmitters and transmission. 2. Television – Transmitters and transmission. 3. Radio broadcasting. 4. Television broadcasting. I. Metcalfe, Shaun. II. Platts, Geoff C. III. Title.
TK6561.W488 1992
621.384′131 – dc20 92–13189 CIP

Typeset by Vision Typesetting, Manchester
Printed in England by Clays Ltd, St Ives plc

Contents

Preface

This book is primarily concerned with radio and television terrestrial broadcasting which, as a means of mass communication, continues to expand in both developed countries and the Third World. In recent years, of course, DBS (Direct Broadcasting by Satellite) has become technically feasible and services are now available in many developed and some Third World countries. However, the advent of DBS most certainly does not signal the demise of terrestrial broadcasting, which offers many desirable features to both the broadcaster and to the viewer or listener. For the broadcaster a terrestrial transmission system is relatively economic to operate and (unlike DBS) specific programmes can be readily beamed to areas with particular cultural and ethnic needs. For the listener and viewer, there is the great attraction of the ready availability of cheap and reliable receiving equipment, and it is no exaggeration to say that the attractions of terrestrial broadcasting are such that it is likely to remain, in the foreseeable future, a primary means of broadcasting worldwide.

There are many books devoted to the principles and techniques of terrestrial broadcasting and their subject matter ranges from the popular to the highly technical. There is, however, little available in the literature that gives practical guidance to the broadcaster concerning the options available to meet his requirements and the factors he must take into account in the selection, design and provision of the optimum broadcasting system for his needs. Large broadcasting organizations, of course, have engineering departments well equipped to deal with these problems, but for other broadcasters, including many in the Third World, it would be quite uneconomic to employ the necessary specialist staff. They therefore have to employ specialist contractors to carry out their technical planning, but at the same time, they have to remain responsible for the scope, cost and successful completion of the work. This book is primarily intended to assist the senior staff of these organizations in the planning, design, installation and operation of terrestrial broadcasting systems, but will be of great interest to newly qualified graduates and others embarking on careers in the transmission side of broadcasting. The aim is not to provide a designer's handbook, since the design of a transmission system cannot be achieved without a large technical staff experienced in the fields of station planning, equipment and system design. Rather, the object is to enable senior staff to make informed decisions concerning the scope of a project and to exercise effective control over the work of specialist contractors, who, as in any field of human endeavour, experience has shown are not infallible. It is hoped that this book will also be of interest to those in the higher managements of larger organizations who are responsible for the final approval of engineering plans and budgets.

The approach adopted is essentially practical, and simple descriptions of methods and equipment are given together with 'rules of thumb' and practical ways of assessing technical and administrative problems illustrated, where necessary, by block schematic figures. Chapter 1 summarizes the necessary steps in the realization of a project from the initial concept to the placing of contracts and their supervision, and Chapters 2, 3 and 4 cover propagation and antennas in the frequency bands from LF to UHF. Chapter 5 covers radio link systems and Chapters 6 and 7 provide a survey of transmitter design and techniques. Chapters 8 and 9 deal with important topics associated with transmission engineering, including civil engineering, computer methods, buildings and structures. Chapter 10 gives a brief summary of the principles and practice of DBS and the development of HDTV (High Definition Television). Chapter 11 summarizes useful information regarding feeders and waveguides and there are two Appendices; one concerning contract administration and the other giving useful definitions, including further explanations of some of the terms used in the text, and formulae. It is impossible to make meaningful technical-planning decisions without some knowledge of engineering fundamentals and the chapters on propagation, antennas, links and transmitters, therefore, include simple explanations of basic principles. However, the text assumes no specialist knowledge: mathematical expressions have been kept to a minimum and used to supplement and quantify the text rather than be its basis.

Chapter 1

General planning considerations

1.1 Introduction

Before considering the technical aspects of a broadcast transmission project it is necessary to review the sequence of events involved in planning and implementation with particular reference to the role of the authority initiating and financing the project (for convenience termed the 'broadcaster'). Those involved in this sequence of events, together with their areas of responsibility, can be listed under four headings:

1. *The broadcaster* who initiates the project, provides the finance and is, in effect, the 'end customer'.
2. *The engineering planners* who are responsible to the broadcaster for the planning and implementation of the project. They are either members of the broadcaster's engineering staff or may be consultants employed by him.
3. *The project manager* who is responsible to the broadcaster for the successful completion of the project. He is either a member of the broadcaster's engineering staff or is provided by the engineering planners.
4. *The contractors* who are employed to carry out the work required by the project. Their work is controlled by the engineering planners with overall co-ordination provided by the project manager.

Although these four areas of responsibility are applicable to all projects, some of them may be combined under one person (or one group of persons). For example, for a simple project involving the provision of a single transmitter, a member of the broadcaster's technical staff could combine the functions of the broadcaster, the engineering planners and the project manager.

When considering the role of the broadcaster it is convenient to divide the sequence of events involved in planning and implementation of a project into the two consecutive phases of *strategy* and *tactics*.

1. *The strategy phase*. Strategy is the art of 'deciding what to do'. The strategy phase, therefore, includes all the studies and activities which lead first to a study of alternative technical plans and finally to the agreement of a detailed technical plan which includes a financial estimate sufficiently accurate for budgeting purposes.
2. *The tactics phase*. Tactics is the art of 'doing what has been decided'. The tactics phase, therefore, follows the strategy phase and includes all the measures necessary to implement the detailed technical plan.

The broadcaster will, of course, be involved in the decision-making process throughout the whole of the strategy and tactics phases. But his greatest involvement is likely to be in the strategy phase and his most important decision will come at the end of this phase. At this point the broadcaster will approve the detailed technical plan for the project, including the financial provision, and if his decision is not correct it may not be possible to modify it, subsequently, without serious financial penalties.

As a guide to a discussion of the strategy and tactics phases, the sequence of events, starting from the initial concept of a project and terminating with the commissioning of the completed installation, is illustrated in schematic form in Figures 1.1(a) and 1.1(b). It should be noted that, although a project may include the provision of studio facilities, this aspect of broadcast engineering is not covered. Studio engineering is a wide subject and it would be impossible to discuss it in detail within the compass of this book.

1.2 The strategy phase

1.2.1 Initiation of a project and schedule of requirements

The starting point of a broadcast transmission project is inevitably social, political or commercial, rather than technical. The need for a new transmitting station, or for a chain of new stations, will arise from the need to provide a new service or, alternatively, to improve an existing service. Initially, the broadcaster must decide, in broad terms, on his objective and then formulate a 'schedule of requirements'. The range of possible objectives is very wide and can extend from the provision of a single transmitter to serve a restricted area to the provision of a complete network of multi-transmitter stations to provide national, regional or area coverage. Alternatively, the objective may be to provide transmissions to overseas countries as opposed to a domestic service. However, experience has shown that significant overall savings, which are often not initially apparent, result if similar requirements (including spares) are ordered simultaneously whenever this is possible rather than at intervals. In the same vein, repeat orders for similar equipment are often preferable to a change to different equipment at a marginally lower cost, as lower spares holdings may be feasible and maintenance staff deployment more cost-effective.

The schedule of requirements is not, primarily, a technical document and need not be couched in technical terms. It should, however, give clear rulings on the following points:

1. The area(s) to be served and the proposed location(s) of the transmitting site(s) (if known).
2. The times, during each 24-hour period, when a service must be provided.
3. The number of programmes to be transmitted.
4. The type of service required. This may range from a simple radio

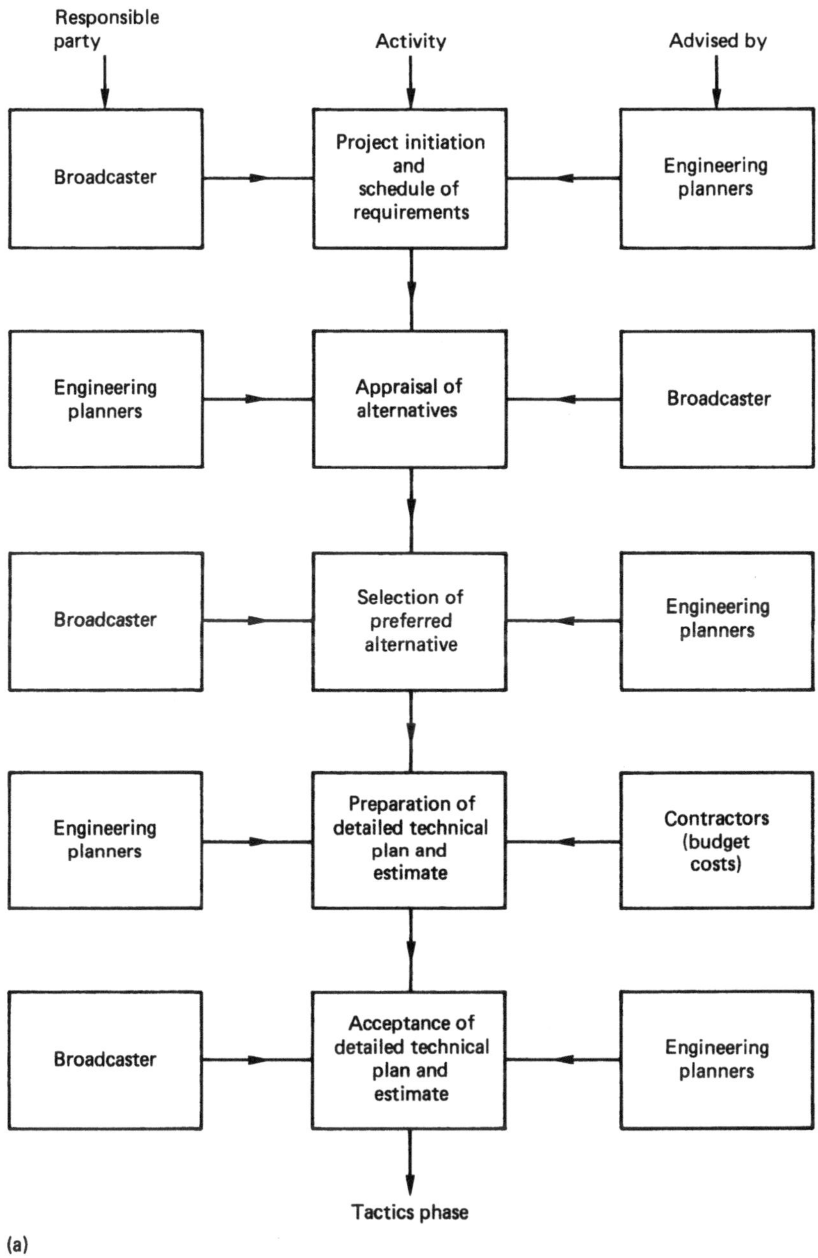

Figure 1.1 Project planning. (a) The strategy phase

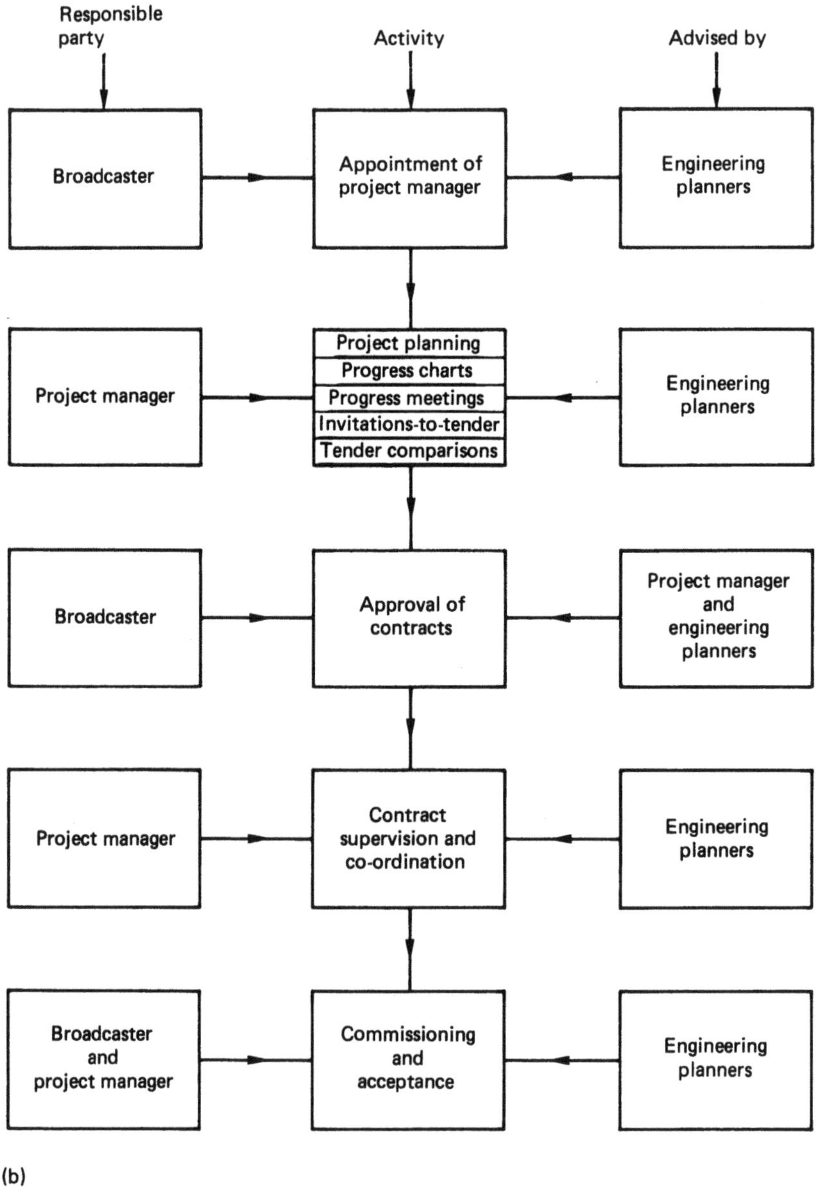

Figure 1.1 Project planning (cont.). (b) The tactics phase

'information service' to a high-quality mono or stereo music service or a colour television service.

5. The target date(s) for provision of the service(s).
6. The minimum acceptable percentage of the population to be served.
7. The approximate financial provision available.
8. The extent to which provision should be made for any future services.
9. Integration (if appropriate) of the new services with existing services, including monitoring, control and maintenance arrangements.
10. Studio facilities requirements.
11. Programme-feed link requirements.
12. If relevant, the need to recruit installation or operating staff.

To illustrate the contents of a schedule of requirements it is useful to consider two examples, one for a radio project and the other for a television project.

(a) Example 1

Schedule of requirements for a national radio service in a small developing country in the tropical zone:

1. The objective is to provide a national radio service as economically as possible and it is essential to provide a service in the three principal towns. The remainder of the country (which has a total area of 30 000 km^2) is sparsely populated and a service, although desirable, is of less importance. There are no existing transmitting sites but ample land is available in the neighbourhood of the three principal towns.
2. A service is to be provided from 0600 to 2359 hours local time but the evening period from 1800 to 2359 hours is particularly important.
3. A single programme will be transmitted but an additional programme may be required at a later date.
4. The programme material will comprise, mainly, news and current affairs.
5. The service is required within 24 months of approval of the detailed technical plan and financial estimate.
6. The whole population of the three towns must be served.
7. A definite figure for available finance is not available but financial economy is of the utmost importance.
8. No programme links are required as on-site studios will be used and telephone lines are available for programme exchanges.
9. Simple studio facilities are required at the transmitter sites.
10. Staff are required for operation and maintenance. Local labour must be used during the installation.

(b) Example 2

Schedule of requirements for a colour television service in a Middle Eastern country:

1. The objective is to provide a colour television transmitting station to serve the capital city and its suburbs which have a total area of 250 km^2.
2. A service is to be provided throughout the whole 24-hour period.
3. A single programme will be transmitted initially but up to two more may be transmitted at some time in the future, and VHF/FM radio coverage with further programmes is expected to be a requirement in the next 10 years.
4. The service will comprise a full range of news and entertainment.
5. The service is required within 18 months of approval of the detailed technical plan and financial estimate.
6. Complete population coverage of the city and suburbs must be achieved.
7. Ample finance is available but the most cost-effective plan will be adopted, but including provision for future services.
8. Studio facilities will form the subject of a separate schedule of requirements.
9. No programme links are required as the studios are to be located at the transmitter site.
10. No operations and maintenance staff are required but local labour is to be used where possible during the installation work.

These two examples are intended to illustrate the degree of detail required in a schedule of requirements and, although typical, do not in any way represent the range of possibilities of transmission-system requirements. Some projects will be technically simple: for example, an increase in transmitter power to improve the grade of service. Other projects, particularly those involving HF overseas broadcasting, will be technically complex. It must be stressed that, although the broadcaster's schedule of requirements is not, essentially, a technical document, it is desirable that technical advice is available during its preparation. It is all too easy to issue a schedule of requirements that prove to be impractical for either technical or financial reasons. In a small broadcasting organization, with no 'in-house' technical planning facilities, the broadcaster will be well advised to retain the services of an engineering consultant to advise on the practicability of the technical, financial and time-scale aspects of the project. However, should technical advice not be available the following guidelines should be of help:

1. The area to be served should, preferably, be illustrated by maps, sufficiently detailed to give an indication of the nature of the terrain (e.g. pastoral, dry, wooded, hilly). It should be borne in mind that the coverage provided by a single transmitting station is limited, however great the transmitter power. The service area radius of a transmitting station depends on the frequency band, terrain, transmitter power and the type of service, and full

information is given in Chapters 2, 3 and 4. However, typical maximum ranges, useful for the preparation of a schedule of requirements, are:

MF Radio	80 km (for frequencies 1000–1650 kHz)
	150 km (for frequencies 520–1000 kHz)*
HF Radio	1000–1500 km (for domestic broadcasting)†
HF Radio†	8000 km (for external broadcasting)
VHF/FM Radio	90 km (for stereo) and 120 km (for mono)
VHF TV	90 km (colour TV)
UHF TV	60 km (colour TV)

In general, for domestic broadcasting, optimum coverage is achieved if the transmitting site is situated on a hill near the centre of the target service area, though not in city centres. However, the lack of main services (mains electricity, water, sewerage, roads and telephone lines) may mean that a centrally located site is not possible. This difficulty can be overcome, to some extent, by the use of a directional antenna radiation pattern.

2. The propagation of MF and HF services varies with the time of day and it is, therefore, helpful to the engineering planners to state the hours during which the service is required and during which good reception is most important.
3. The capital and revenue costs of a broadcasting service increase and frequency allocation problems become more acute as the number of transmitted programmes increases. Careful consideration must therefore be given to the need to transmit more than one programme.
4. The programme quality of a service depends on the type of transmission system and the type of receiver, and ranges from what can be called 'hi-fi' at one extreme to 'transistor-portable' at the other. Hi-fi quality (including the possibility of stereo) is currently only obtainable using VHF/FM (frequency modulation in the VHF band) and is much more expensive to provide than the 'transistor-portable' quality obtainable in the MF and HF bands.
5. The time required for the completion of a broadcast transmission project obviously depends on the technical complexity of the transmission system. As a general guide, the minimum times required for typical transmission projects are approximately:

   (a) Nine months for adding a transmitter or antenna system at an existing site.
   (b) Eighteen months for building a new single-programme transmitting station.
   (c) Three to six years for building a chain of new transmitting stations giving national coverage of a country.

*LF broadcasting (frequencies 153–281 kHz) gives wider coverage but the number of channels is limited and, in general, none are available for new services.

†Although domestic HF (sky-wave) broadcasting gives extensive coverage, reception is subject to noise and fading. It is most suitable, therefore, for speech and light entertainment services.

(d) Three to four years for building a multi-transmitter HF external broadcasting station.
(e) Two to three years for the expansion of an existing HF external broadcasting station with additional transmitters and antennas.

6. It is never possible to guarantee a population coverage of 100% in the service area of a transmitting station. Further, to achieve a population coverage approaching 100% in a large geographical area, with a network of transmitting stations, will be very expensive and may not be technically possible. Particularly for wide-area coverage, it is therefore desirable to state a minimum target coverage either in terms of area or percentage of population. However, for a station serving a single town or city, a population coverage approaching 100% is normally possible and can therefore be specified.
7. It is difficult for the broadcaster to calculate, accurately, the probable cost of his requirements. Even if cost information on earlier projects is available the effects of inflation and of new technology are difficult to assess. However, it is helpful if the broadcaster states the amount of money he is willing to expend or gives some general indication of whether finance is likely to be a problem. The engineering planners will then be able to give costed alternative proposals that take account of the probable financial situation.
8. If studio facilities are required they should be described in broad outline (i.e. number of studios and the types of programme proposed).
9. A radio-link programme feed may be required between the studio and the transmitting station. If this is so, the path distance should be given and, if available, a map showing the terminal locations and the type of terrain provided.
10. Co-siting of LM and MF transmitters with television studios and transmitters is not recommended, as interference to the television services is likely.

1.2.2 The appraisal of alternatives

On receipt of the broadcaster's schedule of requirements, the engineering planners will prepare an 'Appraisal of alternatives' (for convenience referred to as the 'appraisal'). This document will give alternative technical solutions to the problem posed by the schedule of requirements together with estimated time-scales and approximate (budget) costs. Although it is based on detailed technical analyses, the engineering planners should present the appraisal in relatively non-technical terms and, where technical terms are used, their meanings should be made clear. For example, a statement that a particular field strength will be provided in a particular area should be accompanied by an explanation of what this means in terms of the resulting grade of service. The broadcaster should not accept an appraisal that is not comprehensive and clear and which does not provide all the information necessary for a decision concerning the preferred alternative to be reached. To this end, the broadcaster may find the following checklist helpful:

1. If only one technical proposal is made, the reasons why alternative proposals are not presented should be given.
2. For each alternative, if possible, the proposed locations, powers and operating frequencies of transmitting stations should be given. Where relevant, problems concerning frequency allocation and the acquisition of suitable frequencies should be discussed.
3. The predicted coverage, provided by the alternative technical proposals, should be given, preferably in the form of maps with contours showing the limits of first and second class service areas. Clear definitions of the standard of service corresponding to the first and second class service areas should be given. In the case of MF proposals, any difference between day and night coverage should be shown. In considering coverage it is necessary to distinguish between 'programme quality' and 'grade of service'. Programme quality describes the faithfulness of reproduction of the received programme and, as has already been pointed out, can be conveniently classed as 'hi-fi' or 'transistor-portable'. Grade of service refers to the incidence of interference: a first class service is virtually interference-free and a second class service is subject to a level of interference likely to be acceptable to the listener or viewer.
4. The availability of main services (electricity, sewerage, water supply, road access and telephone lines) must be discussed and any difficulties made clear.
5. Any difference between the equipment reliability provided by alternative proposals should be explained. This can be in terms of MTBF (the mean time between failures).
6. Where relevant, the advantages and disadvantages of alternative types of building should be discussed.
7. The approximate (budget) costs for each alternative should be given, preferably broken down under the following headings:

   (a) Transmitters
   (b) Masts, antennas and transmission lines
   (c) Power supplies
   (d) Buildings (including furniture and fittings)
   (e) Civil engineering works: roads, hard standing, fencing, etc.
   (f) Installation and commissioning of technical equipment
   (g) Annual maintenance costs
   (h) Annual staff costs (where relevant)
   (i) Annual operating costs

   Cost figures will normally be given at 'today's prices' and, if the project is likely to occupy two years or more, it may be necessary to make some allowance for inflation. This is not a straightforward matter and the broadcaster may need to take expert advice on the probable effect of inflation on future capital and revenue costs.

8. Additional cost of making reasonable provision for future services – this may be limited to ensuring that the tower or mast can support any necessary additional antennas and that the site size is adequate to accommodate additional transmitter buildings.
9. The estimated time-scale for the completion of each alternative proposal must be given. These estimates will normally be given in weeks or months from the date on which the project is authorized to proceed.
10. If called for in the schedule of requirements, the problems likely to be encountered in the recruitment of qualified operating and maintenance staff should be discussed and any need for further training by qualified staff (quite often expected) made clear.

1.2.3 Discussion of alternatives

On receipt of the appraisal, the broadcaster will hold detailed discussions with the engineering planners. These meetings should be chaired by a senior member of the broadcaster's staff and the appraisal should be discussed on a paragraph by paragraph basis. It may be possible for the broadcaster to select a preferred alternative but it may transpire that the costs of all the alternatives exceed the available funds. The engineering planners will then be asked to produce lower cost, 'scaled-down', proposals. Scaling-down can be achieved in a number of ways but, in general, it is likely to involve a reduction in coverage, quality or grade of service, or technical reliability. If a reduction in coverage, or in quality or grade of service, is proposed the broadcaster must decide whether the resulting service will be acceptable. Further, to a lesser extent a reduction in technical reliability (i.e. a greater frequency of breakdowns) may affect the acceptability of a broadcasting system. Thus the engineering planners will provide the technical information on which a scaling-down decision can be based but the final decision must, of course, be made by the broadcaster.

In an area where no service (or no comparable service) exists, it may well be that a second class service will be successful in attracting and satisfying listeners or viewers. However, if the proposed service is to be in competition with an existing service, it must be at least equal (and preferably better) in quality and reliability. This is because listeners and viewers will tune to the strongest, and most reliable, signal that provides the type of programme that they want. It follows that a service that would be successful, if operated in isolation, may fail to attract listeners or viewers in competition with an existing service with a higher field strength.

The engineering planners will base their predicted coverage on a minimum field strength that will be provided in the service area. It is necessary, however, for the broadcaster to ensure that the engineering planners make clear the quality and grade of the service that will be provided and, if in any doubt, the broadcaster can require a practical demonstration. Such a demonstration can be arranged with relatively simple equipment and should take place in an area (or areas) where the field strength is expected to be a minimum and man-made electrical noise is expected to be a maximum. For these tests, a typical receiver

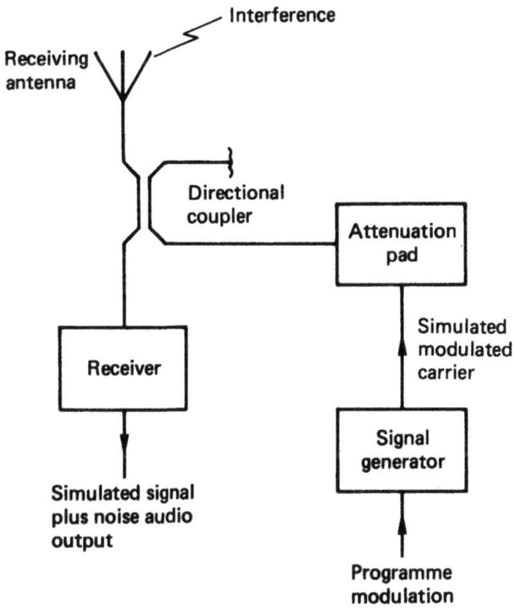

Figure 1.2 Received signal quality: test arrangements

together with a typical domestic receiving antenna is used. A signal generator (in effect a very low power transmitter) is connected to the receiver and arranged to inject the predicted input-signal level using programme modulation typical of that of the proposed service. The arrangement is illustrated in Figure 1.2 and will enable the quality of the service to be assessed and, where necessary, compared with that of a competing service.

1.2.4 Example of selection of the preferred alternative

When the appraisal of alternatives has been discussed, and where necessary modified, the broadcaster will proceed to the selection of the preferred alternative. In the foregoing discussion, an attempt has been made to draw attention to the points that the broadcaster must bear in mind when initiating a project and when considering the appraisal of alternatives prepared by the engineering planners. These points can best be underlined by an example and it is convenient to base this on Example 1 of a schedule of requirements given earlier in this chapter (requirement for a national radio service in the tropical zone). In presenting an appraisal of alternatives for this example it would be misleading to give costs correct at the time of writing as, due to inflation, they would become rapidly out-of-date. To avoid this difficulty, but at the same time to enable costs comparisons to be made, prices are given in 'units' chosen so that the cost of a 10-kW MF transmitter is 100 units. The appraisal of alternatives, produced by the engineering planners in response to the broadcaster's schedule of requirements, could be in the following form.

(a) General technical considerations

Since financial economy is important, only one programme is to be transmitted and, as the programme content is to be mainly speech, the programme quality need not be in the 'hi-fi' category. As a result, only relatively basic broadcasting systems, operating in the MF and HF bands, need be considered as they are significantly cheaper in terms of capital and revenue cost per unit area of coverage than are VHF/FM systems. The programme quality provided by VHF/FM systems is superior to that of MF and HF systems but is not necessary for the type of service proposed. The choice, therefore, lies between an MF system which will provide a good grade of 'transistor portable' service and an HF system which will be cheaper but provide a lower grade of service. Another possibility is a service comprising a combination of MF and HF transmissions.

(b) Coverage of an MF system

Daytime MF transmissions are by 'ground-wave' and the coverage depends on the transmitter power, the frequency and the conductivity of the terrain. During the hours of darkness the ground-wave is still present but there is also a sky-wave formed by energy reflected from the ionosphere. Where the amplitude of the sky-wave is comparable with that of the ground-wave, there is an area of severe fading and distortion, known as the 'fade-zone', where reception is virtually impossible. The fade-zone may therefore reduce the coverage during the hours of darkness. The terrain of the country is dry and is estimated to have a relatively low conductivity of 5 mS/m (see Chapter 2). Also, the three frequencies allocated by international agreement are in the higher part of the MF band (960, 1200 and 1400 kHz). Assuming a transmitter power of 10 kW and an antenna height of 0.25 wavelengths (giving omnidirectional coverage), the predicted coverages for the three allocated frequencies are as shown in Table 1.1.

It will be seen that, for all three frequencies, the fade-zone lies beyond the limit of the second class service area so that there will be no reduction of coverage at night. If a transmitting station is sited at, or near, each of the three principal towns, a first class service will be provided in each town with a second class service to a number of rural communities.

(c) Coverage of an HF system

It is possible to provide a broadcasting service over a wide area using a single HF transmitter and directing the signal vertically (a system known as 'sky-wave

Table 1.1

	Ground-wave service area radius (km)		
Frequency (kHz)	*1st class service**	*2nd class service*†	*Radius of fade zone (km)*
960	39	59	109–162
1200	31	44	100–140
1400	26	38	90–130

*1st class service with minimal interference.
†2nd class service with acceptable interference.

broadcasting'). After reflection from the ionosphere, the signal returns to earth and can serve an area up to 100 times that of a single MF transmitter. The system is economic in terms of cost per head of population served but the service provided can only be regarded as second class. The reason for this is that the ionosphere is not a stable reflector and the received signal is therefore subject to fading. Further, the characteristics of the ionosphere vary with the diurnal, seasonal and 11-year sunspot cycles so that more than one frequency is necessary to ensure an adequate received signal at all times. Thus the listener will need to re-tune his receiver from time to time and, at certain times in the sunspot cycle, he may need to tune to different frequencies for day and night reception.

Calculations show that, if a 100-kW HF transmitter and sky-wave transmitting antenna are located near the principal town (i.e. roughly in the centre of the country), an effective service will be provided over the whole country. Three frequencies of approximately 7.3, 5.4 and 3.4 MHz will be needed to cover the diurnal, seasonal and sunspot ionospheric changes.

(d) Frequency allocations

The predicted coverage for the MF and HF services, given in (b) and (c) above, takes account of calculated co- and adjacent-channel interference at the assumed frequencies. It will be necessary for the broadcaster to arrange for his PTT – the Government ministry whose responsibilities include those of frequency planning and allocations – to seek formal approval for the use of the assumed frequencies, and the proposed transmitter powers.

(e) Site services

For both the MF and HF proposals it will be possible to select transmitting sites that are adjacent to mains electricity, water supplies and telephone lines. There is no main sewerage but the cesspit arrangements, in common use locally, can be used. At all sites, access is available from local roads that are, in no case, more than 300 m from the edge of the site.

(f) Loss of service due to technical failures

Well-engineered, and highly reliable, antennas are commercially available for the proposed MF and HF services. The MTBF (mean time between failures) of these antennas is well in excess of five years and repairs are not likely to be time-consuming, providing inspections are made to the antenna systems at yearly intervals. It is therefore considered that standby antennas need not be provided. In the case of the MF service it is proposed that transmitters of an all-solid-state design are used (solid state transmitter powers of up to 200 kW are now available). The multiple-transistor output stage of this type of transmitter provides 'built-in redundancy' and therefore gives high reliability and obviates the need for a standby transmitter. In the case of the HF system an output power of 100 kW is required and since an all-solid-state design with this high power is not available, it will be necessary to use a transmitter with a tube (tetrode) output stage. Modern tubed transmitters, although less reliable than all-solid-state designs, have a relatively high MTBF and short repair times and it is therefore not considered necessary to provide a standby transmitter. It should also be noted that, as discussed in Section 8.3.4, a good earthing and lightning protection system is essential for good reliability and that this is particularly true in the case of solid-state transmitters.

The mains supply at all the sites can only be regarded as moderately reliable and it is therefore recommended that diesel-alternator standby generators, with automatic changeover, are provided.

(g) Buildings

Since financial economy is of prime importance it is recommended that an 'industrial' type of building is used. These buildings are robust, cheap to maintain and have a life expectancy of at least 30 years. None of the proposed sites is in a 'prestige' area and the 'factory' appearance of industrial buildings should be quite acceptable.

(h) Studio equipment

Studio facilities are required and this appraisal includes the minimum equipment for the transmission of a single programme. At each site allowance is made for a single talks studio, a continuity studio and minimum tape, turntable and programme input equipment (the equipment necessary to feed the transmitter). The talks studio will accommodate discussion programmes with up to six participants and would also be suitable for small drama productions.

The MF proposal

Capital costs should be checked thoroughly and listed as shown, for example, in Table 1.2.

Table 1.2

Item	*Quantity*	*Total cost*
Mast radiators and feeders	3	120
10-kW transmitters	3	300
Power supplies (including standby)	3	90
Buildings, furniture and fittings	3	120
Civil engineering works	3	150
Studio facilities	—	60
Installation and commissioning	—	200
Contingency	—	60
Total budget cost		1100

The annual revenue costs should also be estimated to be 500 units which includes power, services (water etc.), spares consumption and staff costs. This is known as 'annual cost of ownership'.

The HF proposal

Again capital costs should be detailed, as shown, for example, in Table 1.3.

Table 1.3

Item	*Quantity*	*Total cost*
Antenna and feeder	1	200
100-kW transmitter	1	1000
10-kW standby transmitter	1	200
Power supplies (including standby)	—	30
Buildings, furniture and fittings	—	40
Civil engineering works	—	50
Studio facilities	—	20
Installation and commissioning	—	70
Contingency	—	30
Total budget cost	—	1640

The annual revenue costs also need to be estimated to be 200 units which include power, services (water etc.), spares consumption and staff costs.

(i) Budget costs

The prices of broadcast equipment are continually increasing and, due to the unpredictability of inflation, it is impossible to give a formula to adjust prices to the correct value on any given date. The main purpose of giving budget costs is to enable alternative proposals to be compared and, since the relative costs are unlikely to change, the conclusions concerning the proposals should remain valid.

(j) Timetable

The time-scale for completion of the MF and HF proposals will be similar and will depend, primarily, on the delivery of the transmitter and antenna equipment. Bearing this in mind the time-scale is estimated to be as shown in Table 1.4.

Table 1.4

Tender enquiries	3 months
Placing of contracts	2 months
Delivery of technical equipment	9 months
Installation	4 months
Commissioning and acceptance	3 months
Total time for completion	21 months

It should be noted that the total project completion time of 21 months dates from approval of the detailed technical plan and a simultaneous availability of funds. It is anticipated that the funds will be expended in three equal amounts phased over three successive periods of three months.

(k) Recruitment of staff

The budget costs for the installation and commission of the equipment assume that the necessary specialist staff are provided by the contractors and that both skilled and unskilled labour is recruited locally. So far as the operations and maintenance of the stations is concerned, it is assumed that two fully trained and experienced staff will be recruited to act as chief engineer and senior maintenance engineer. Technically educated local technician staff should not be difficult to recruit and can be trained in their duties by the specialist engineers. It should be possible, after a period of two or three years, for sufficient trained local staff to be available to take over complete running of the stations. This will obviously result in a saving in revenue costs but it is recommended that an estimate of the saving is made when the detailed estimate for the project is prepared.

(l) Summary of alternatives

There are three alternative proposals for the provision of a national radio service:

1. A network of three MF stations that will provide a first class service in the three principal towns and a second class service to surrounding communities. A service will be provided to about 80% of the national population but the remaining 20% will remain unserved.
2. A single HF station that will provide a second class service to virtually the whole population of the country. The capital cost of this station will be little more than half the cost of the network of three MF stations.
3. In this proposal, alternatives (1) and (2) combined could be provided. This would provide very comprehensive coverage of the country and the residents in the towns would have an alternative programme source in the event of a failure of the MF service. It would also be possible to contemplate the transmission of two different programmes at certain times of the day.

If finance is an overriding consideration an attractive solution would be to opt, initially, for alternative (2) (a single HF station). This would provide comprehensive coverage of the whole country in a relatively short time and it would be possible to add MF transmitters to improve the service in the towns if further finance became available.

1.2.5 The detailed plan and estimate

When the broadcaster has decided on his preferred alternative he will instruct the engineering planners to prepare a detailed technical plan together with a firm financial estimate. This will be an expanded version of the chosen alternative and must satisfy the following requirements:

1. Any modifications agreed in the discussions with the broadcaster must be included.
2. A full description of the equipment and facilities (including any for future development) to be provided should be given. Where possible manufacturer's brochures should be included.
3. The costs given should be based on the response to enquiries to typical manufacturers.
4. A 'ceiling' figure for cost should be given by the addition of a contingency sum (say 10–15%).

The broadcaster must study the detailed plan and estimate in great detail as he has reached a 'point of no return'. He must be quite clear concerning the population coverage, programme quality, grade of service and transmission reliability that will result from the proposed financial expenditure. He must also be clear on the revenue costs of the proposed system and, if necessary, take expert financial advice on the probable effects of inflation. In his capital budget the broadcaster will be wise to allow an additional sum of 10% per annum to allow for inflation and unforeseen price increases.

1.2.6 Acceptance of the detailed plan and estimate

Acceptance of the detailed technical plan and estimate by the broadcaster brings to an end the strategy phase of a project. The acceptance implies a mandate for the engineering planners to proceed with the project and to commence to expend the agreed funds. Further, the broadcaster has made a decision that will be difficult to modify, or cancel, without serious financial penalties. Except for simple projects, authority to proceed should be in the form of a protocol document, issued by the broadcaster, defining the scope of the work to be carried out, the total funds agreed and the rate at which these funds can be expended. The form of words used will depend, to some extent, on the nature of the project but the following outline gives a guide:

1. Authority is given for the work proposed in the detailed technical plan and estimate (reference number of the plan and estimate, together with the reference numbers of any amendment documents) to proceed. The following provisions are included in this approval.
2. No technical and financial changes to the agreed plan are to be made without written authority. In particular, any changes affecting coverage or financial expenditure must be submitted for approval.
3. A project manager will be appointed to act on our (i.e. the broadcaster's behalf) and will be responsible for the timely completion of the project within the allocated funds and at the agreed expenditure rate.
4. The project manager will submit regular progress reports to us (i.e. to the broadcaster) covering technical and financial progress.
5. All contracts for a sum in excess of (an amount decided by the broadcaster) will be submitted for our approval.
6. The total sum allocated at today's prices is (insert sum) and the expenditure of these funds is to take place at the following rate (insert the funds available in each three-monthly period). Each contract submitted for acceptance by contractors must make clear any liability for cost rises due to inflation or exchange rate variations.
7. A statement should accompany each contract submitted for approval and should summarize all tenders received and give reasons for the selection of the chosen tender.
8. The date by which the project is to be completed and service will commence is (insert date). Any potential delay affecting the completion date must be reported to us (i.e. to the broadcaster) via the project manager.

The preparation of this document should be discussed between the broadcaster and the engineering planners as it commits both parties to a clearly defined course of action. The broadcaster is committed to the expenditure of funds and the engineering planners are committed to the construction of a system providing a specified coverage at an agreed cost.

1.3 The tactics phase

With the acceptance of the detailed plan and estimate, and the authorization to the engineering planners to proceed with the project, the most important decision of the broadcaster has been made. However, as will be seen from the acceptance document, outlined above, the broadcaster has an important role to play in monitoring the technical and financial progress of the project via the project manager. It might be argued that the implementation of a project involves work of a purely engineering nature and that the engineering planners should be allowed to proceed without interference from the relatively non-technical broadcaster. Nevertheless, whilst it is true that the broadcaster has little part to play in the field of purely technical decision-making, he is the 'end customer' for the project and must be in a position to intervene if it appears that engineering problems are likely to endanger its successful completion either technically or financially. Here the importance of the project manager is clear since it will be his duty to apprise the broadcaster of any problems likely to occur. The broadcaster will then initiate discussions with the engineering planners to decide the best way of solving the problem.

1.3.1 Appointment of the project manager

It must be stressed that it is difficult to bring any project, other than a very simple one, to a successful conclusion without the appointment of a project manager responsible for monitoring all aspects of its implementation. The staff employed on the implementation of a project will cover a wide range of disciplines and the project manager cannot be responsible for making specialist decisions in all the fields involved. The engineering planners must therefore be responsible for their own professional decisions and the role of the project manager must be that of a co-ordinator. He must be an engineer with sufficiently broad experience to understand in general terms the specialized work of the planning staff and to appreciate the nature of the problems that will inevitably arise. He is not responsible for the actual control of the project but for monitoring progress on all aspects of the work. In particular, he is responsible for drawing the attention of the broadcaster to potential delays, overspending and failures to meet agreed technical standards.

The project manager is appointed, jointly, by the engineering planners and by the broadcaster, to whom he will be responsible for reporting progress. The duties of the project manager can be summarized as follows:

1. To be fully conversant with the detailed plan and estimate for the project and maintain close contact with the engineering planners and the contractors responsible for the various aspects of the work.
2. To maintain regularly updated progress charts for the project.
3. To co-ordinate the preparation of all invitations-to-tender and to ensure that these are completed and contracts are issued in accordance with the agreed timetable.
4. To hold regular progress meetings at a frequency agreed with the broadcaster and the engineering planners.
5. To issue a progress report, covering both technical and financial aspects of the project, subsequent to each progress meeting.
6. To report promptly to the broadcaster any reason, or reasons, for delay or over-expenditure.
7. To resolve any problems arising in the co-ordination of the work.

It will be seen that the project manager, although not a line manager, bears a heavy responsibility and plays an essential role in the successful completion of a project. The project manager will normally be a member of the engineering planning organization and may, in a small project, undertake his duties in addition to his normal work. For large projects it is desirable to appoint a full-time project manager, and for very large projects it may be necessary to appoint an additional consulting engineer with experience of project management for the duration of the project.

1.3.2 Project planning and invitations-to-tender

The first task of the project manager will be to study the detailed plan and estimate prepared by the engineering planners and to draw up a main progress chart that will be the basis of the implementation and co-ordination of the project. For very large projects this may need to be in the form of a computer-based spreadsheet or a critical-path-analysis (CPA) chart. It must be pointed out, however, that a CPA chart does not give a simple presentation of the progress of a project and its main purpose is to enable the effects of difficulties occurring in a complex project to be readily evaluated. When presenting progress information to the broadcaster and the engineering planners it will be necessary for the project manager to prepare simple, and easily understood, summaries in written or bar-chart form which can be derived, where necessary, from a CPA chart or computer-based spreadsheet.

When the main progress chart has been drawn up and agreed with the engineering planners, the project manager will present it to the broadcaster for his agreement. So far as the broadcaster is concerned, completion of the main progress chart signals that an essential step in the implementation of the project has been completed. The main progress chart is the master document for the conduct of the project and will be updated regularly so that it can form the basis for the agenda for the regular progress meetings held by the project manager. An example of a main progress bar chart is given in Figure 1.3 for the HF Radio Service Project used to illustrate the preparation of a schedule of requirements and summary of alternatives in Section 1.2.4. It will be seen that the chart illustrates the essential steps in the implementation of a project and, in particular, includes dates for:

1. The completion and issuing of all invitations-to-tender.
2. The receipt of all competitive tenders.
3. The completion of comparison-of-tender documents and selection of chosen tenders.
4. The placing of all major contracts.
5. The commencement and completion of all work on site.
6. The commissioning, testing and putting into service of all equipment.

The project manager should ensure that each invitation-to-tender is issued to at least two or preferably three or four contractors but there is little point in issuing invitations-to-tender to a large number of contractors. Contractors whose facilities and qualifications for carrying out the work successfully are in doubt should not be invited to tender: it wastes their time and leads to difficulties when tenders are assessed. It should also be noted that it is not necessary to issue detailed specifications for items of equipment that are manufactured in a standard form. For example, a transmitter manufacturer will offer a range of transmitters of standard design and specific power outputs. The manufacturer may be willing to modify his standard equipment to accommodate local conditions; for example, for a different mains supply voltage, for an extended ambient temperature range or to provide increased air-filtering for a dusty environment. But he is unlikely to be willing to make fundamental technical changes aimed at, for example, increasing the efficiency or extending the operating frequency range. The transmitter specification, therefore, need only cover the essential points of power output, operating frequency and environmental conditions. But information concerning the efficiency and 'lifetime ownership costs' should be sought.

The work involved in the preparation and issue of invitations-to-tender is of a purely technical nature and does not need to be supervised by the broadcaster. This work is carried out by the engineering planners and co-ordinated by the project manager, who will report progress to the broadcaster.

1.3.3 Comparisons of tenders and placing of contracts

The preparation of comparison-of-tenders documents is the responsibility of the project manager based on information provided by the engineering planners. These documents will compare the competitive tenders received, both technically and financially, together with a recommendation as to the tender to be accepted. On the basis of these documents the broadcaster will be asked to

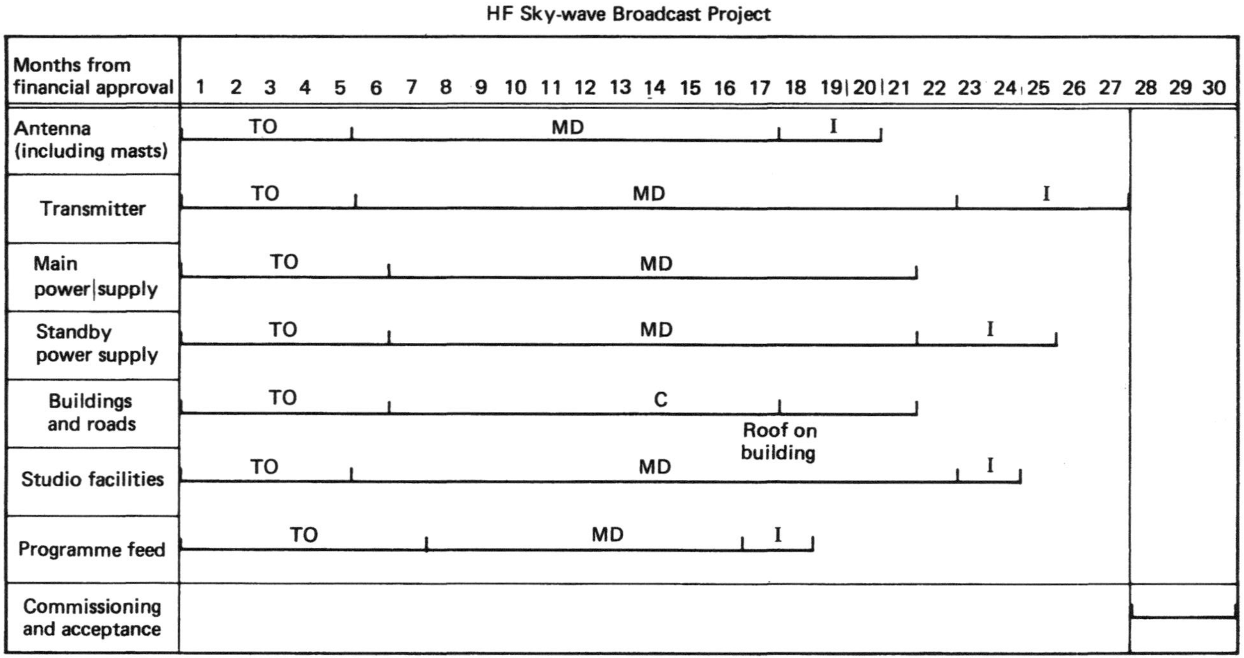

Figure 1.3 Typical progress bar-chart. TO, tendering and order; MD, manufacture and delivery; I, installation and testing; C, construction

approve the award of a contract to the recommended contractor. It must be pointed out that a comparison of competing tenders may not be straightforward. This is because the equipment offered may differ technically, or because equipment offered by one tenderer is in service with other broadcasters and is thus a 'proven design' (with the manufacturer's contracted delivery date more likely to be met) as against the latest 'drawing board' designs of other tenderers which may be subject to development delays and subsequent 'teething troubles', or because tenders are not always presented on the same financial basis. When equipment is offered which has already been in service with another broadcaster, a visit to that broadcaster to ascertain his experience with the equipment is often illuminating. One contractor may offer a fixed price for the work whilst another may base his offer on a formula that makes allowance for rises in labour and material costs due to inflation. It is the responsibility of the project manager to ensure that competing tenders are summarized and compared on an equitable basis in each comparison-of-tenders document. The document will recommend the tender that is considered the 'best buy' and give clear reasons for the choice. The project manager must ensure that all necessary information is included and that the conclusions drawn are presented in a logical and clear manner. A broadcaster with limited engineering resources may well be advised to let someone else experience the 'teething troubles' of a new 'automated' design, particularly if his operating and maintenance staff costs are likely to be relatively low and their engineering experience limited.

1.4 Contracts

The most economic way to provide a transmitting installation is to place separate contracts for the major items of equipment and the major parts of the work. For example, separate contracts can be placed for transmitters, masts, antennas, buildings and civil engineering. By this means, the 'best buy' can be achieved in each area of expenditure with a consequent reduction in the total cost of the project. Although an attractive method of contract administration, the placing of separate contracts is most suited to large broadcasting organizations who have the specialist staff to be responsible for the overall project planning and to co-ordinate the work of a number of contractors. Thus a broadcaster with no 'in-house' engineering planning facilities of his own may decide that he does not wish to or indeed cannot exercise detailed technical responsibility and that the project should be carried out on a 'turnkey' basis. In this case a single contractor will be employed to carry out all the work, employing sub-contractors as necessary, and to offer a complete and working installation for acceptance. An intermediate approach may be for the broadcaster to employ a contractor to provide the antenna to meet the area coverage requirements and its support structure, a different contractor for the remainder of the technical equipment and a third one to provide the accommodation specified by the technical equipment supplier. However, the broadcaster will still need to retain an engineering consultant to protect his interests and to carry out the functions of the engineering planners and the project manager that have been described.

The types of contract available, and their administration and financial implications, are fairly complex matters and are explained in detail in Appendix 1.

1.5 Transmitter network strategy

1.5.1 Formulation of policy

Whenever the origination and transmission of a new radio or television service is contemplated it is prudent to consider the likely ramifications as part of a decision to proceed. Defined policies and planning for the future at this stage can, often at little extra cost, result in substantial savings in the long term.

Clearly, factors of importance in particular cases will differ, but are likely to be greatly influenced by the broadcaster's experience, if any, of existing services and the existence of ongoing development of transmitter networks for those services, and of course available finance. However, as a guide, some of the questions which need to be asked to formulate policies when new services are contemplated are considered below and some of them are dealt with in greater depth in later sections of the book.

1.5.2 Reliability and continuity of service

It is important to distinguish between reliability of service and reliability of equipment. The former is the reliability as perceived by the listener or viewer, whilst the latter determines the maintenance cost – costs of spare parts and the employment of maintenance staff.

In an ideal world all viewers and listeners would enjoy the same standard of service, but in practice economic considerations do not permit this. As a general rule, high-power transmitting stations:

1. although the most costly to build, serve proportionally higher numbers of people than low-power stations.
2. are more complex than low-power stations, with high-power stages more prone to failure.

For these reasons high- and medium-power transmitting stations are often equipped with duplicated transmitters which normally operate in parallel, or reserve transmitters with automatic changeover (and sometimes automatic tuning when one transmitter is a reserve for several services), with powers of only a tenth that of the main transmitter. Due to automatic gain controls in receivers, such an arrangement leads to seriously reduced standards of service only on the fringes of transmitter service areas.

Clearly, the parallel arrangement, if well engineered, will provide a very reliable service, but maintenance effort and costs associated with two transmitters and their paralleling equipment can be expected to be almost doubled relative to one transmitter.

The system employing automatically switched reserve transmitters is less reliable as there is no guarantee that either the changeover equipment or transmitters will be serviceable when required.

Faults on antennas, particularly on tall structures and during inclement weather, may take a considerable time to repair so that, even if duplicated feeders are not necessary for power-carrying reasons, it is prudent to use such an arrangement at high-power, high-population-coverage stations.

Whilst it is desirable that all transmitting equipment is highly reliable, this will be to no avail if the programme fails or mains supplies are unreliable.

1.5.3 Network and service expansion

Typically a new service will be provided by a high-power transmitting station to serve the main centre of population with other high-, medium- and low-power transmitters subsequently built and commissioned to serve other areas. In these circumstances the first transmitting station should be designed and sited, so far as this is practicable, with future developments in mind; for example, it might be feasible to site it so that its location is also suitable for a link site for a programme feed to the next transmitter to be built in the chain or for it to have a directional radiation pattern so that it would be received at the proposed location of other transmitters.

If expansion of services is envisaged, antenna systems should be designed with adequate power-carrying capacity and bandwidth for the future developments – costs are by no means proportional to power-carrying capacity. Similarly, structures should have the strength to carry additional antennas.

As networks and services expand it may be necessary to improve the quality and reliability of interconnecting programme feeds, and to fit antenna systems with over power protection circuits.

1.5.4 Network control

A network of transmitters can be manually controlled, remotely controlled or automatically controlled.

Originally all transmitters were manually controlled and maintained by on-site staff, but with increasing reliability and the extension of networks of relatively few high-power transmitting stations, first with more medium power and subsequently many low-power ones, manual control of all the stations was no longer economically feasible, or necessary, or indeed desirable from the aspect of job satisfaction.

A network of transmitters, e.g. one providing national coverage of a particular programme, is now usually engineered so that start-up of the first one is initiated by a programme from the originating studio and subsequent ones are initiated in a chain reaction. This is achieved by means of receivers which detect the presence of transmissions by the preceding stations. However, at the lowest power levels, say up to 10 W, it is often considered more economic to leave the transmittters permanently powered rather than fit switch-on equipment with its attendant complication, risk of failure and maintenance demands.

Sometimes there is a need for remotely controlled transmitters, e.g. if it is desired to use only part of the network or use different parts for different programmes. When such systems are used care must be taken to ensure that remote control systems cannot override any local protection systems. Experience is that there is a tendency for operational staff, presumably in their desire to restore service, to consider that protection systems have operated spuriously and to continue to override them, resulting in serious damage.

1.5.5 Monitoring, maintenance and spares philosophy

Some monitoring systems enable many parameters of transmitting station equipment to be interrogated from a central location, but a preferable arrangement is for monitoring devices on transmitting stations to signal the central monitoring location when a fault has occurred. The essential information for maintenance staff is either 'Come as quickly as you can; the programme is not being transmitted or is a grossly distorted version of the incoming feed signal' or 'Come as soon as convenient; the programme is being transmitted by reserve equipment, at low power or as a somewhat distorted version of the incoming feed signal.' Some additional information, e.g. to indicate operation of an intruder alarm or to advise on spares which may be required, may be worth the additional complexity of the monitoring system, but anything which can be classified as merely of interest and will not speed repairs should be avoided. The information may be signalled back to the central monitoring location either by telephone link, radio link or by direct monitoring of the transmissions, or a combination of these.

The cost and complexity of monitoring automatically controlled inherently reliable low-power transmitters serving smaller communities in this way may not be justified and arrangements with a member(s) of the public to report faults have often been found to be satisfactory in practice. Indeed, if several programmes are being transmitted it is feasible to design competitively priced equipment so that a report of the service or services which have failed indicates the nature of the fault.

Increasingly with modern automatically controlled transmitters, either

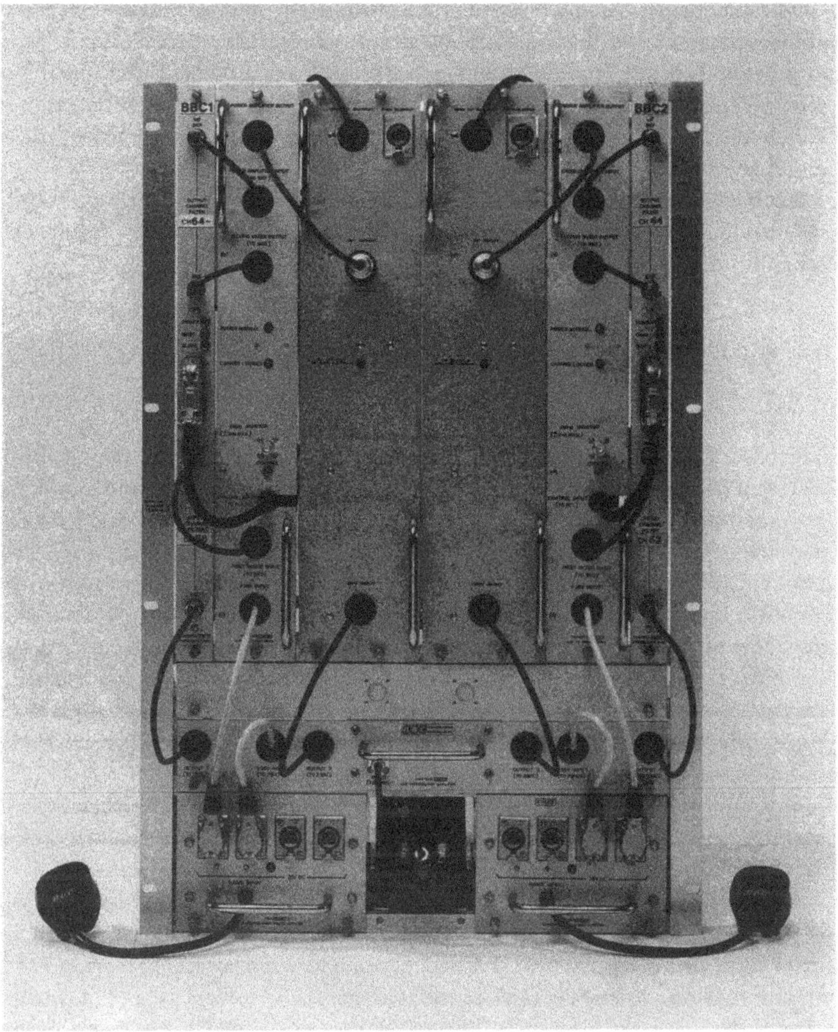

Figure 1.4 A two-programme × 10 W UHF/UHF modular construction transposer assembly. The 10-W wideband amplifier modules occupy the same space as a 2-W transposer and its passive channel determining 'personality panel', thus permitting the modules to be configured in a variety of ways, e.g. four programme × 2 W or four programme × 10 W (with an additional frame to carry the 10-W amplifiers) (Courtesy of Continental Microwave, Luton)

duplicated or in association with reserve transmitters, maintenance staff are mobile and may be directed by radio to wherever their services are required and it is generally feasible to work 'office hours'. Faults signalled to monitoring centres 'outside office hours' can be recorded for action during the next working day.

Modern practice for a network of transmitters is to use similar types of equipment whenever this is possible and to stock spares on an area basis. More and more transmitting equipment is being built on a modular basis, so that relatively unskilled staff can locate and replace faulty modules, a selection of which are carried in their vehicles, returning the faulty ones to centralized maintenance workshops.

Mobile and centrally based maintenance staff will note the nature of faults, which will be recorded and subsequently analysed to determine equipment reliabilities and initiate any desirable modifications.

1.5.6 Suitability and tenure of sites

The cost of acquiring a site is effectively not only the cost of the site itself, but also that of providing the services (electricity, communications, and water and sewerage facilities if requested) and access, and there may be charges if these pass over land owned by others. It is therefore prudent to negotiate for all the requirements simultaneously to avoid the risk of being held to ransom at the eleventh hour by any of the parties involved. If the site can only be leased, adequate security of tenure and terms for renewal of lease should be agreed at the outset. Any agreement should be for 'use of the site for radio frequency transmissions' rather than for, say, 'a television service', as the latter gives the site owner the possibility of additional fees if the site is subsequently required for radio or communications or programme link transmissions.

The site should be large enough for any building extensions which may be required and planning permission, if required, should be obtained before agreements to purchase are entered into.

It is generally not feasible to use the same site for LF, MF or HF transmission as for VHF or UHF transmissions, as the frequencies and harmonics of the former fall within the video or IF frequency bands of the latter. Furthermore, MF and LF sites usually use energized insulated masts and towers as antennas which cannot then be used to carry VHF or UHF antennas, whilst HF stations have many masts or towers to support curtain antennas which would reflect VHF and UHF transmissions, distorting radiation patterns and giving rise to delayed signals.

1.6 Safety and risks to personnel

1.6.1 Introduction

Every broadcast transmission authority has a duty of care towards its staff, contractors, the general public and visitors to its premises. Conversely contractors and others have a duty to protect the broadcaster's staff, and all should be aware of the need to comply with any national health and safety at work legislation. However, no instruction, certificate, permit or other document can in itself give complete protection. A considered approach together with previous 'in house' or 'consultants' experience of similar projects is the best way to minimize the risk of accidents, but it should be appreciated from the outset that to achieve a high standard of safety does not come free of charge. Nevertheless such charges may well be less than the cost of litigation or damages for injury and the cost of repairing damaged plant, quite apart from the consequences and distress of personal injury or death.

1.6.2 General hazards and those peculiar to transmitting stations

The general hazards found on transmitting stations are similar to those found in large light-engineering establishments or in office environments and are not dealt with in this book except to note that adequate numbers of suitable strategically located fire extinguishers, together with smoke detectors, fire alarms, fire doors and exit notices, should be provided.

Hazards peculiar to transmitting stations arise from the presence of masts and towers, RF radiations, high voltages and, to a lesser extent, from lower voltage low-impedance supplies, standby generating plant, toxic substances and high-voltage capacitors.

The dangers associated with high structures may appear to be obvious: the danger of personnel falling or being hit by something dropped from above. Less obvious is the risk of lightning strikes, whilst RF radiation and high voltages cannot be directly perceived.

Hazards during the construction of a transmitting station or subsequent modifications or re-engineering are likely to be very different, and involve different people to those employed to operate and maintain completed stations, and need to be carefully considered.

1.6.3 Management of risk

(a) General

A broadcasting authority should have a formalized approach to health and safety at work and appoint an officer to formulate, document and be responsible for such matters, including regular reviews of and updating of documentation.

A new or expanding authority is likely to benefit from the experience of the larger national broadcasting authorities, who may well be willing to act as consultants.

When a number of contractors or the broadcaster's own staff are simultaneously on a site, one party, which may be the broadcaster, must be appointed to be responsible for safety on the site and his responsibilities and authority clearly defined, and the other parties advised accordingly. At the tendering stage and during any pre-contract discussions, contractors must be acquainted with the extent of their responsibilities and duties in respect of safety.

Safety officers should organize safety courses appropriate to the needs of the various categories of staff, including refresher courses. Subjects covered might include fire evacuation procedures, first aid, including resuscitation techniques, and work on high structures,

All accidents and near misses should be reported and recorded, and details circulated within the organization as considered appropriate. First aid kits should be prominently located. Procedures to be followed in an emergency should be defined in advance and publicized by notices, e.g. resuscitation after electric shock, telephone numbers of local police, medical and fire-fighting services.

Meeting agenda should always include safety as an item. To do so draws the attention of those attending meetings to the importance of safety and encourages them to discuss incidents.

Modern transmitters are likely in themselves to be satisfactorily interlocked to prevent access to live components, but care should be exercised if different contractors are used to supply the antenna system, power supply or cooling plant to ensure that there are no interface problems affecting safety, particularly if transmitters are remotely controlled. Guards should be fitted over all moving parts and lockable local isolators fitted to remote ancillary equipment, e.g. heat exchanger fans or antenna/load switches. Lock and interlock keys should be held by those carrying out maintenance work so that equipment cannot be re-powered until the keys have been returned.

Protective clothing, including goggles and ear defenders, should be available and staff obliged to use them when appropriate, e.g. when grinding metals or in close proximity to operational diesel generators.

Although batteries and low-voltage low-impedance supplies are not usually a hazard so far as electric shock is concerned, they may be able to provide very heavy short-circuit currents, sometimes explosively melting the short-circuiting conductor, resulting in injury or a fire. Secondary batteries can release explosive gas mixtures during charging; consequently they should be housed in well-ventilated rooms.

High-voltage capacitors should be stored with their terminals short-circuited. Capacitors discharged by momentarily short-circuiting their terminals can become recharged by residual charge retained in the dielectric.

Strategically positioned notices should warn against specific hazards or advise of necessary safety procedures.

Equipment containing toxic substances should be labelled accordingly and the procedure for the disposal of such substances made known to maintenance staff.

Power supplies for general use on transmitting sites should be fitted with sensitive (30 mA) earth leakage circuit breakers and portable tools should preferably operate at 110 V from centre-tapped earthed secondary isolating transformers.

Carbon dioxide fire-extinguishing systems, often installed in unmanned rooms containing oil-filled transformers and switchgear, should be fitted with disabling controls and warning indications of the state of the system to avoid the risk of suffocation in the event of operation whilst someone is in the room.

Staff visiting unmanned automatic transmitting stations should preferably work in pairs and so far as is practicable advise of their arrival on site, and confirm their presence at defined intervals and when they leave site. Ideally they should be in radio contact with their base.

(b) Masts and towers

All personnel required to climb and work on high structures or operate hoists or winches must undergo training before so doing. In addition they should be passed as medically fit at regular intervals by a doctor.

Personnel should be equipped with protective clothing, hard hats and safety harnesses (not a simple waist belt) with an easy-to-operate means of attachment to structures. If a greater range of movement is required, inertial reel blocks should be used with the harness. Protective clothing should have pockets designed to minimize the risk of things being dropped.

Material and man-carrying hoists should be load-tested at the start of each day. Man-carrying hoists should always be railed and operated with inertial fall arresters running on twin taut guide wires, so that in the event of a winch or bond failure the hoist platform would not fall more than one or two metres – personnel should, of course, ensure that they attach themselves to the hoist platform by means of the safety harness.

Masts and towers should always be equipped with ladders and rest platforms at intervals of about 10 m. Working platforms should have sheet rather than open mesh floors, with raised edges to minimize the risk of objects being knocked off.

Masts and towers should be fitted with anti-climbing devices and warning notices. It is suggested that notices with wording which threatens potential unauthorized climbers may be a more effective deterrent, e.g. 'Danger to Life: RF Radiation', rather than 'Climbing Prohibited'.

Power supplies on masts and towers for socket outlets should be fed from isolating transformers with secondary windings of 110 V centre tapped to earth.

Aircraft obstruction lights, with a minimum of three at each level, shall be fitted if and as required by the responsible authority. Flashing xenon gas discharge lamps are to be preferred over filament lamps due to their long life and high intensity, and should be wired on two circuits for reliability.

The erection of masts and towers should always proceed with an eye on the weather to ensure that potentially hazardous operations can be completed before the onset of adverse weather and that partially completed structures will be able to withstand a change to similar conditions. Temporary stays (guy ropes) must be used as necessary and be adequate to carry loads imposed by the design specification.

When work is in progress aloft the area in which objects may fall should be delineated and access to it prohibited.

Work should not take place if lightning is thought likely and towers should always be earthed before an ascent is made.

(c) Radiation hazards

The possibility of radiation hazards should be considered when a transmitting station is designed and steps taken to ensure that neither operational staff nor the public will be exposed to such risks during the normal operation of the station. A survey of levels of radiation should be carried out on completion of the station in proximity to both the transmitters and antenna and feeder systems.

The most likely risk of exposure will be to personnel carrying out maintenance work or installing new or re-engineering existing plant, particularly on masts carrying several services. Radiation hazard meters should be supplied for use in these circumstances and their calibration checked at regular intervals. This may be rather time-consuming if calibration facilities are not easily available. For this reason it is a good idea to have two meters which can be checked against each other and for repeatability of a predetermined reading made at a precisely located position, with the meter in a defined orientation, where there is a safe level of radiation produced by a transmitter operating at a known power level.

Information about safe levels of radiation are given in Section A2.1 and on Figure A2.8.

Chapter 2

LF and MF propagation and antennas

2.1 Introduction

In the years after World War II, it was thought that LF and MF broadcasting might be phased out and replaced by frequency modulation in the VHF band (VHF/FM). The limited number of LF and MF channels available, and the consequent co-channel interference, degrades night-time reception so that the more expensive VHF/FM system (which does not suffer from this disadvantage) seemed to offer a very attractive alternative. This prediction has proved quite wrong and, in the past 40 years, there has been a steady increase in the use of MF broadcasting in parallel with an increase in VHF/FM broadcasting. However, the very limited number of channels available has prevented any corresponding increase in the use of the LF band. There are a number of reasons for the continued use of MF as a popular radio broadcasting medium in both industrialized countries and in the Third World:

1. The development of solid-state techniques has made cheap and easy-to-tune LF and MF domestic receivers readily available.
2. MF broadcasting, particularly during the daytime, provides greater coverage, and at a lower cost per head of population, than does VHF/FM broadcasting.
3. The worldwide demand for broadcasting channels makes it impractical to give up the LF and MF bands.

2.2 Propagation

To provide a compact and convenient measurement scale, LF and MF field strength is usually given in dBu (decibels above 1 μV/m) and, for example, a field strength of 10 mV/m (10 000 μV/m) is given as 80 dBμ (see Section A2.1). The field radiated by an LF or MF antenna system, over flat perfectly conducting ground, decreases inversely as the distance at all angles of elevation. Thus, for example, if the field strength at a distance of 20 km from a transmitter is 50 mV/m (94 dBμ) then the field strength at a distance of 40 km will be 25 mV/m (88 dBμ). In practice, this 'inverse distance law' is true for radiation at angles above the horizontal but the field in the vicinity of the ground decreases much more rapidly with distance because of ground losses (due to finite ground conductivity) and the curvature of the earth. It is convenient, therefore, to regard the field radiated by a practical LF or MF antenna as comprising two components; the 'ground-wave' and the 'sky-wave'.

2.2.1 The ground-wave

The ground-wave is the field in the vicinity of the ground and is used for LF and MF domestic broadcasting. The ground-wave is vertically polarized and can only be generated by a vertical 'monopole' antenna (a vertical wire, or alternatively, a mast) energized by connecting a transmitter at its base. The ground-wave field strength provided by a vertical antenna decreases with increasing frequency and with decreasing ground conductivity, which both have a profound effect on propagation. To illustrate this, Table 2.2 shows the distance at which a field strength of 60 dBμ (acceptable for MF rural reception as shown in Table 2.3) is obtained under the conditions shown in Table 2.1.

Table 2.1

*Effective radiated power (ERP)**	1 kW
Frequencies	500, 1000 and 1600 kHz (the bottom, centre and top of MF band)
Ground conductivities	1, 5 and 10 mS/m which correspond, respectively, to 'poor', 'average' and 'good' conductivity ground

*The product of transmitter power and antenna gain relative to a short (less than 0.1 wavelengths high) monopole which is accepted as the 'standard' antenna (see Section A2.3).

Table 2.2

	Ground conductivity (mS/m)		
Frequency (kHz)	*1*	*5*	*10*
500	39 km	87 km	117 km
1000	19 km	44 km	62 km
1600	12 km	27 km	38 km

Table 2.2 shows that, with the lowest frequency (500 kHz) and highest conductivity (10 mS/m), the range is 117 km, whereas with the highest frequency (1600 kHz) and the lowest conductivity (1 mS/m), the range is reduced to 12 km (a reduction of around 10: 1).

2.2.2 The sky-wave

The sky-wave is the field radiated by an LF or MF antenna at angles above the horizontal and is of importance as it can be reflected back to earth by the 'ionosphere', which comprises layers of ionized gas created by ultraviolet radiation from the sun. During daylight hours, the effect of the ionosphere can be ignored, as the sky-wave is absorbed by the 'D-layer'. At night, however, the D-layer vanishes and the sky-wave is reflected back to earth by either the E-layer or F-layer, depending on frequency. The E- and F-layers exist, respectively, at heights of about 100 and 200 km above the earth. The process by which the sky-wave is returned to earth is, actually, one of refraction but, for convenience, it is common practice to describe it as reflection.

Due to ionospheric changes, the sky-wave reflected back to earth at night varies in both magnitude and phase and, in the area (known as the 'fading zone') where its amplitude is comparable to that of the ground-wave, fading and distortion of the received signal occur. Depending on the radiated power, antenna height, frequency and ground conductivity, the incidence of fading can reduce the service area during the hours of darkness and this can best be illustrated by an example. If the radiated power is 50 kW, the transmitted frequency is 550 kHz and the ground conductivity is 5 mS/m (average terrain), a ground-wave field strength of 60 dBμ (the minimum acceptable for a rural service) occurs at a distance of 180 km. However, during the hours of darkness, fading occurs over the range of distances from 100 to 170 km (quarter-wave radiator) and, as a result, the night-time ground-wave service area radius is reduced to 100 km from its daytime value of 180 km. It should be noted that, during the hours of darkness, the sky-wave from a co-channel station may also cause interference and limit the coverage of an LF or MF transmitting station.

2.2.3 LF and MF reception

The ground-wave field strength required to provide a service depends on the level of natural and man-made noise and the loss of field strength due to shadowing by natural and man-made obstructions in the reception area. For simplicity, however, methods of calculating ground-wave and sky-wave field strength assume that the signal is propagated over a smooth, unobstructed earth. For planning purposes, field strength is calculated on a smooth-earth basis and the required value is chosen to allow for natural and man-made noise, field strength variation and the effect of obstructions. Taking FCC (Federal Communications Commission) and CCIR (*Comité Consultatif International des Radio Communications* – International Radio Consultative Committee) recommendations into account, Table 2.3 gives minimum calculated smooth-earth field strengths that give acceptable reception in practice.

Table 2.3

Reception area	*LF (dBμ)*	*MF (dBμ)*
Rural areas[a]	72	60
Rural areas[b]	78	66
Suburban area	85	73
City areas[c]	92	80
City areas[d]	99	87

[a]Isolated rural dwellings.
[b]Small villages and suburbs of towns.
[c]Towns.
[d]Large cities and industrial areas.

These minimum field strengths will give a good standard of reception but can be reduced by 6 dBμ if a marginal standard of reception is acceptable. The two standards are sometimes referred to as first class reception (field strengths in Table 2.3) and second class reception (field strengths in Table 2.3 minus 6 dB).

The values of field strengths in Table 2.3 apply to the centres of the LF and MF bands (218 kHz and 1000 kHz respectively) and can be used for initial station planning. For a more accurate assessment of station coverage the field strengths can be adjusted by using the correction factors (interpolating as necessary) given in Table 2.4.

Table 2.4

Band	*Frequency (kHz)*	*Correction (dBμ)*
LF	155	+1.1
	218	0.0
	281	−1.1
MF	500	+5.5
	750	+2.5
	1000	0.0
	1300	−2.0
	1600	−3.1

So far it has been assumed that reception is not subject to interference from co-channel or adjacent-channel stations. When this occurs, in order to suppress audible interference, the wanted field strength must exceed the unwanted interfering field strength by an amount known as the 'protection ratio'. Useful values of protection ratio are:

1. *Co-channel interference.* Interference by signal with different programme: 30 dB. Interference within a synchronized network: 8 dB. (A group of synchronized co-channel stations carrying the same programme.)
2. *Adjacent-channel interference.* Interference by stable signal (9 kHz bandwidth): 9 dB. (This protection ratio assumes a limited degree of modulation compression (often used on speech and light entertainment transmissions) on the wanted and unwanted transmissions.)

The minimum field strength required to achieve interference-free reception is known as the 'protected field strength'. For example, if there is a co-channel interfering field strength of 35 dBμ then the protected field strength will be $35+30=65$ dBμ.

2.2.4 Propagation analysis

The object of propagation analysis is to predict the service area of a proposed broadcast transmitting station.

(a) Ground-wave field strength

There will be a required minimum smooth-earth ground-wave field strength (see Tables 2.3 and 2.4) and this will depend on the frequency band, the noise level at the receiving point, obstructions, and the required grade of service. Propagation analysis, therefore, involves a comparison of calculated smooth-earth field strength with the required minimum smooth-earth field strength (which, of course, depends on the reception area environment).

A precise calculation of LF or MF smooth-earth ground-wave field strength requires a complete solution of the fundamental electromagnetic equations. However, a relatively simple formula is available that gives the field strength with sufficient accuracy for virtually all practical purposes and has been used by the CCIR and other authorities as a basis for standard propagation curves. With this formula, the field strength in dBμ is given by the sum of a number of factors each expressed in dB:

1. The field strength in dBμ given by a short vertical monopole antenna, with an ERP (effective radiated power – see Section A2.8 (Appendix 2)) of 1 kW, assuming that the path is flat and perfectly conducting (see Section A2.3 (Appendix 2)).
2. The ERP actually used expressed in dB above 1 kW (this is 10 log (ERP)) where the ERP is in kW).
3. The attenuation due to the conductivity of the ground.
4. The attenuation due to diffraction round the curvature of the earth.

The sum of these factors gives the field strength on the assumption that the ground conductivity has a uniform value over the entire path. When this is not so, a semi-empirical method due to Millington can be used in which the path is divided into sections in which the conductivity can be considered constant. This method can be illustrated by the case of a two-section path in which 100 km of high-conductivity land is followed by 50 km of low-conductivity land. The field strength at a distance of 100 km over high-conductivity ground is first calculated. The loss over the distance from 100 to 150 km over poor-conductivity ground is then calculated and deducted from this first value to give the field strength at 150 km. The calculation is repeated for the opposite direction of propagation (i.e. for a path comprising 50 km of low-conductivity land followed by 100 km of high-conductivity land) and the two calculated field strengths are averaged. The averaging process is necessary as the field strength calculated in the two sections will, in general, differ. Useful values for the conductivity and permittivity of different types of terrain are given in Table 2.5.

Table 2.5

Terrain	*Permittivity*	*Conductivity (mS/m)*
Sea	80	4000
Fresh water lakes	80	1
Swampy land	30	20
Cultivated land	10	14
Average terrain	4	5
Forest terrain	13	4
Rocky terrain	14	2
Dry sandy terrain	10	1

Calculations show that ground irregularity does not have a significant effect on received ground-wave field strength unless the terrain is very hilly. Even at the upper end of the MF band (1600 kHz), the maximum to minimum heights of the undulations of the ground must exceed about 200 m (600 ft) before they need to be taken into account. However, ground-waves suffer additional attenuation when they pass through urban (city) areas. It is found that as a signal passes through a built-up area it is attenuated but, at distances beyond the area, there is a recovery in field strength. This attenuation is known as 'urban loss' and, as a guide, is about 5 dB for suburban areas and 10 dB for city areas. These losses are taken into account in the required minimum smooth-earth, ground-wave, field strengths given in Table 2.3.

(b) Sky-wave field strength

Sky-waves originate at a transmitting antenna and terminate at a receiving antenna after reflection by the ionosphere. As they travel over this path, they are attenuated and this attenuation comprises three components defined as the 'ground loss', the 'polarization coupling loss' and the 'ionospheric absorption loss':

1. The ground loss is due to loss of energy in the ground (which has finite conductivity) in the region of the transmitting and receiving antennas. This loss can be regarded as constant (except for sites near the sea).
2. The earth's magnetic field has a considerable influence on the propagation of the sky-wave. The free electrons in the ionosphere are set in motion by the transmitted wave but their paths are deflected due to the presence of the earth's magnetic field. This deflection of the electrons reaches a maximum at a frequency around 1400 kHz which is known as the 'gyromagnetic frequency' (conveniently abbreviated to 'gyro frequency'). At frequencies in the region of the gyro frequency (i.e. in the MF band) waves in the ionosphere are elliptically polarized and can only be excited efficiently by an incident elliptically polarized wave. However, MF antennas radiate vertically polarized waves and thus will not efficiently excite an elliptically polarized wave in the ionosphere. Also, the elliptically polarized wave emerging from the ionosphere will not fully excite the listener's receiving antenna (normally only sensitive to vertical polarization) effectively. These losses (on entry to the ionosphere and at the receive antenna) are called 'polarization coupling loss' but only become significant in tropical latitudes and then only on long east–west transmission paths.
3. When a transmission passes through the ionosphere, the free electrons are set in motion and, in some cases, collide with gas molecules, thus causing a loss of energy and consequent 'ionospheric absorption loss'. At low frequencies the electron movement is large and the number of collisions is large so that the absorption is high. As the frequency is increased, the absorption is reduced and the wave returned to earth is increased in amplitude. However, eventually a frequency is reached at which the ionosphere becomes transparent so that the wave is not returned to earth. In the MF band, low frequencies are reflected by the E-layer (at a height of 100 km) but at frequencies above 1 MHz the E-layer is usually penetrated and reflection is from the F-layer (at a height of about 200 km).

A formula for sky-wave field strength prediction is recommended by the CCIR and applies to path lengths up to 12 000 km. The formula gives the median field strength (i.e. the average field strength of a fading signal) in dBμ and comprises the sum of a number of factors, each of which is expressed in dB:

1. *ERP* (effective radiated power, which combines transmitter power and antenna gain – see Section A2.8 (Appendix 2)) expressed in dB above 1 kW, which is given by 10 log (ERP) where the ERP is in kilowatts.
2. *Path loss* (attenuation due to the length of the path assuming free-space propagation with no losses).
3. *Ionospheric loss* (attenuation due to absorption in the ionosphere which is dependent on frequency, path length and latitude). This loss includes an allowance for the polarization coupling loss occurring in low latitudes. It

also includes an allowance for transmitting and receiving ground loss in areas not adjacent to the sea.

4. *Sea gain* (the increase in received signal which occurs when the transmitter or receiver is situated near the sea). Sea gain is significant only for path lengths greater than 1000 km and when the transmitter or receiver is located less than 5 km from the sea.
5. *Excess polarization coupling loss* (the amount by which the polarization coupling loss exceeds the allowance made in the ionospheric loss above). This loss occurs in the MF band but not in the LF band, is greatest in the east–west direction, but can be ignored at latitudes above 30°. Further, at latitudes below 30°, it is only significant in directions of propagation within ±20 degrees of magnetic east–west.

Summarizing the above; sea gain and excess polarization coupling loss do not, in general, have a significant effect on the location of the fading zone of an MF transmitting station. As has been pointed out, a fading zone will exist at night due to interaction of the sky-wave and ground-wave and will take the form of an annulus (a ring) surrounding the station in which reception is impossible. The fading zone will occur in the area where the amplitude of the sky-wave, and that of the ground-wave, differ by 6–10 dB or less and may limit the night-time coverage. Beyond the fading zone the sky-wave will predominate and can provide a high field strength at distances very much greater than those covered by the ground-wave. It is possible, therefore, to serve a very extensive area during the hours of darkness providing fading due to ionospheric variations is acceptable.

2.3 Antennas

In the HF, VHF and UHF bands, broadcast requirements can, generally, be met with standard types of transmitting antenna available from a wide range of manufacturers. However, this does not apply in the LF and MF broadcast bands where the requirements as regards operating frequency, radiation pattern, transmitter power, radiation efficiency, mast height and coverage vary from station to station. Thus, in general, standard antennas are not used in LF and MF broadcast bands and each case is considered on its merits. However, the need for a specific design for each application is offset by the fact that an LF or MF antenna system can be assembled from relatively simple standard parts (masts, insulators, wires and impedance matching components).

2.3.1 LF and MF radiator types

The simplest LF or MF broadcast antenna comprises a single vertical monopole antenna, mounted on a base insulator. The monopole is driven by a transmitter connected across the insulator and provides an omnidirectional horizontal radiation pattern (i.e. it radiates the same field in all directions). In most cases, the monopole takes the form of a stayed mast radiator with the stays broken up by insulators to make them electrically transparent, as illustrated in Figure 2.1(a). Alternatively, a self-supporting tower, as illustrated in Figure 2.1(b), can be used in cases where the transmitting site area is restricted. It is also possible to earth the base of the monopole and connect the transmitter to a point on the mast as shown in Figure 2.1(c). This arrangement, known as a 'shunt feed', avoids the need for a base insulator and is often used in areas subject to a high incidence of lightning strikes. In all cases a ground mat, comprising a number of buried radial wires, is used to minimize power losses due to finite earth conductivity. The ground mat is illustrated in Figure 2.1(d) and it will be seen that one side of the transmitter output is connected to the mast, or tower, and the other to the mat.

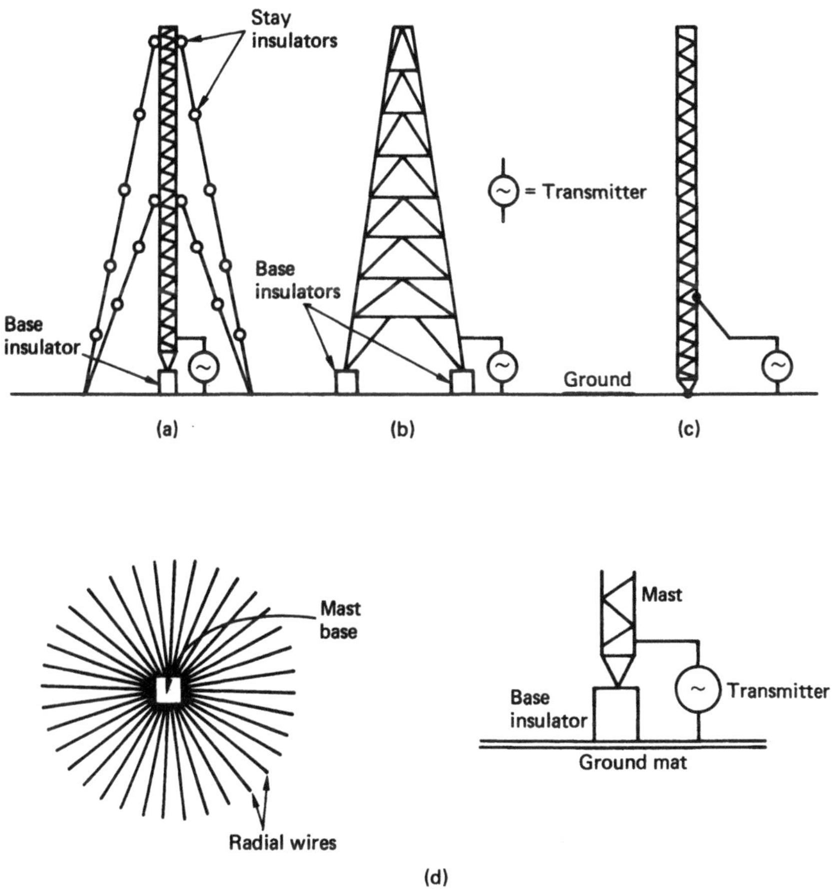

Figure 2.1 LF/MF broadcast antennas. (a) Mast radiator. (b) Tower radiator. (c) Shunt-fed mast radiator. (d) Radial-wire ground mat

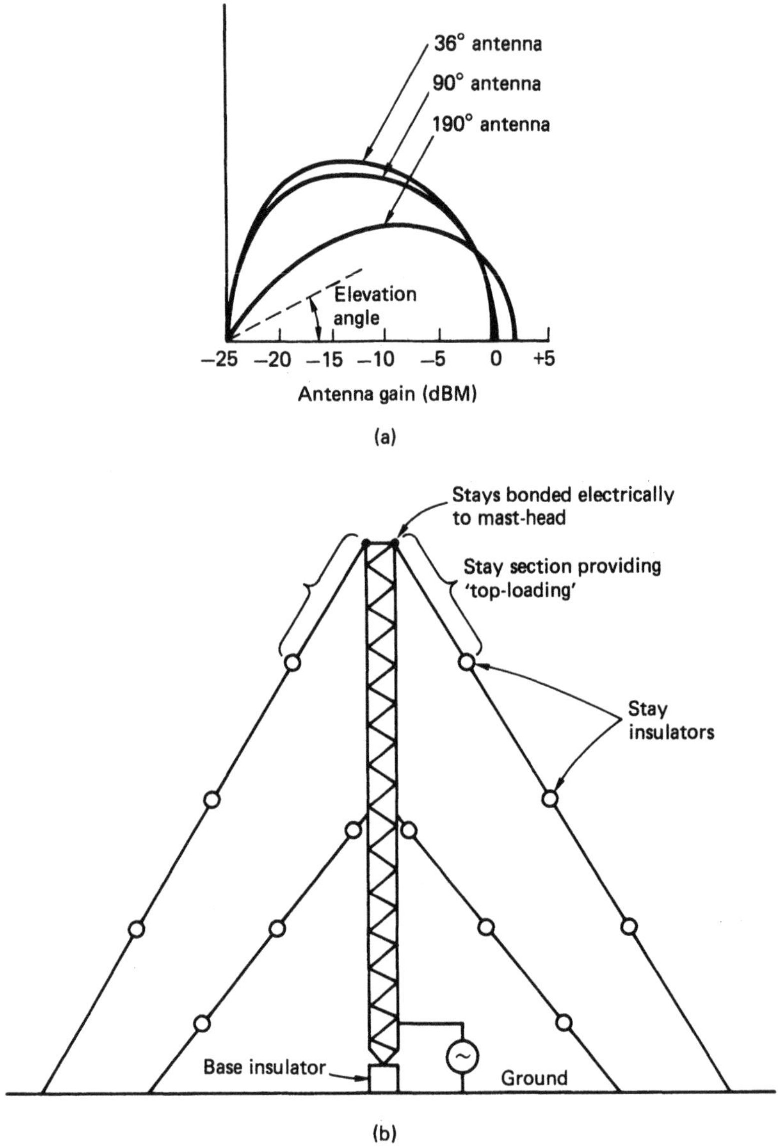

Figure 2.2 (a) Vertical radiation pattern, LF/MF antennas. (b) Top-loaded mast radiator

The vertical radiation pattern (VRP) of an LF or MF antenna depends on the electrical height and, in practice, varies from about 0.1 wavelengths (36°) to about 0.55 wavelengths (200°). The VRP changes little with heights up to a quarter-wavelength but, as the height is further increased to a half-wavelength and above, the radiation above the horizontal is significantly reduced, as illustrated in Figure 2.2(a). As a result, the fading zone is moved further from the transmitter, and monopoles with a height in the region of a half-wavelength are, therefore, known as 'anti-fading' antennas. It should be noted, however, that there is no point in providing an expensive anti-fading antenna unless the night-time coverage is limited by fading.

In the LF band, and at the lower frequencies in the MF band, an anti-fading antenna will require a very high, and expensive, structure (for example, at 520 kHz the height will be about 300 m (1000 ft)). It is possible, however, to achieve most of the advantages of an anti-fading antenna by using a shorter structure and increasing its electrical height by a 'top-loading', as illustrated in Figure 2.2(b). In this case the top-loading is achieved by connecting, electrically, the upper sections of the stays to the mast-head.

2.3.2 Gain and input impedance

The gain of an antenna has been mentioned in connection with the calculation of ground-wave and sky-wave field strength and is explained in Section A2.3. Briefly, gain in the LF and MF bands is defined as the ratio of the transmitter powers required to provide the same field strength from the antenna under consideration and from a standard short monopole. Thus the gain G is given by:

$$G = 10 \log (W_s/W_A) + 20 \log (E_A/E_S) \text{ dBM}$$

where W_A is the power radiated by the antenna, W_s is the power radiated by the standard monopole, E_A is the field stength provided at the antenna, and E_S is the field strength provided by the monopole at the same distance. The 'M' in dBM denotes that the gain is relative to a short monopole and, based on this definition, the gains of typical omnidirectional LF and MF antennas of typical heights are: up to 0.1 wavelengths, 0.0 dBM; 0.25 wavelengths, 0.4 dBM; 0.55 wavelengths, 2.0 dBM.

An antenna radiates power and the input impedance (sometimes called the drive-point impedance) must therefore include a resistive term which is known as the 'radiation resistance'. The current on a monopole is distributed approximately sinusoidally, as shown in Figure 2.3(a), so that the antenna behaves, approximately, as a uniform open-circuited transmission line. It is convenient to assume that the radiation resistance is equivalent to a single resistance connected in series with the antenna at the current maximum, which is referred to as the 'loop current'. Thus defined, the radiation resistance is called the 'loop radiation resistance'. Using the fact that the antenna behaves, approximately, as a uniform transmission line, the equivalent circuit of a monopole antenna is shown in Figure 2.3(b). The equivalent circuit can be used for the approximate calculation of drive-point impedance but is not sufficiently accurate to give the input impedance of an antenna comprising two or more monopoles. In these cases it is necessary to use a computer program based on a solution of the fundamental electromagnetic equations and described in Chapter 9.

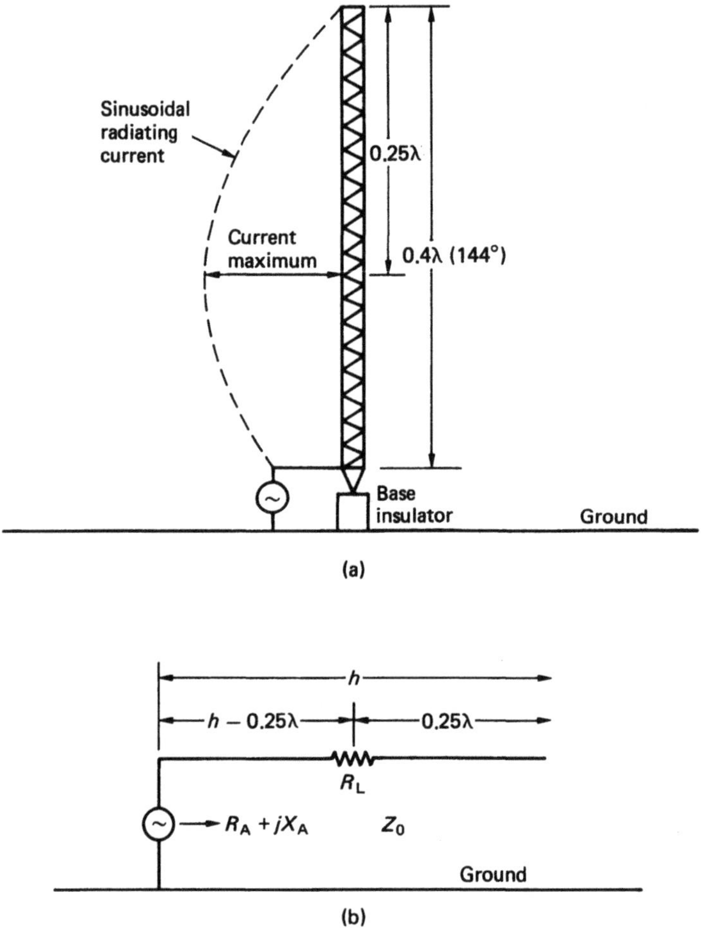

Figure 2.3 Current distribution and antenna input impedance. (a) Example of sinusoidal radiating current on LF/MF monopole. (b) Equivalent circuit of antenna. h, height of antenna (monopole) in wavelengths; R_L, loop radiation resistance; Z_o, $138 \log (h/a - 1)$, antenna characteristic impedance; a, radius of antenna; $R_A + jX_A$, impedance at antenna base

2.3.3 Radiation efficiency

The 'radiation efficiency' of an antenna is a measure of the reduction of radiated power due to losses in the antenna and in the feeder system. The principal cause of reduction of radiated power is the dissipation of energy in the ground in the vicinity of the antenna and this depends on the ground conductivity and the size and density of the ground mat. Radiation efficiency is defined as the ratio of the radiated power and the input transmitter power, expressed as a percentage. However, this definition, although apparently straightforward, is difficult to interpret in practice as it involves an arbitrary distinction between the losses occurring in the ground surrounding the antenna (which obviously decrease efficiency) and those associated with the attenuation in the propagation path (which obviously do not). A simple approach to the problem is to define 100% efficiency as that occurring with a lossless ground mat (e.g. a circular copper sheet) with a radius of one wavelength and then to calculate the energy absorbed within one wavelength with the actual ground mat used. This approach is known as the 'absorbed energy method' and can lead to an efficiency exceeding 100% if a dense earth mat with a radius exceeding the wavelength is used. However, ground mats with a radius exceeding one wavelength are very seldom used in practice, but the absorbed energy method can be regarded as giving a useful indication of efficiency. Typical values of radiation efficiency, calculated by the absorbed energy method, are given in Table 2.6.

Table 2.6

Height of antenna (wavelengths)	*Radius of ground mat (wavelengths)*	*Number of radials in ground mat*	*Radiation efficiency (%)**	
			1 mS/m	*10 mS/m*
0.1	0.05	4	32	56
0.1	0.1	36	64	82
	0.1	72	66	85
0.25	0.1	36	73	89
0.25	0.25	36	80	91
0.25	0.25	72	82	93
0.25	0.25	120	83	94
0.55	0.1	72	84	95
0.55	0.25	72	85	95
0.55	0.5	72	88	96
0.55	0.5	120	89	96

*Conductivities of 1 and 10 mS/m correspond to ground of poor and good ground conductivity respectively.

Table 2.6 shows that radiation efficiency changes relatively slowly with the number of radial wires and their length. The minimum radiation efficiency acceptable in practice depends on transmitter power and coverage. For 1-kW power installations an efficiency as low as 60% (−2 dB) can be acceptable but for high-power installations an efficiency of less than 85% (−0.7 dB) will represent an uneconomic loss of transmitter power. For high-power installations (50 kW and above) a radial ground mat with 120 wires (3° apart) and a radius of a quarter-wavelength is generally used and offers a reasonable compromise between cost and efficiency.

2.3.4 Antenna matching and VSWR

LF and MF transmitters are designed to work into a resistive load that is chosen to be equal to the characteristic impedance of a commercially available feeder (normally either 50 Ω unbalanced or 300 Ω balanced). To provide the correct resistive load for the transmitter, therefore, the antenna impedance at the carrier frequency must be transformed by a matching network so that it is equal to the characteristic impedance of the feeder. In practice, the input impedance of an LF or MF antenna will be very different from the characteristic impedance of the feeder so that a matching network must provide a large impedance transformation. The accuracy of matching is measured in terms of the VSWR (voltage-standing-wave ratio defined in Section A2.5 (Appendix 2)) on the feeder between the transmitter and antenna, which is 1:1 for a perfect match and 1:0 for a complete mismatch. In a typical case, the antenna must be matched to the characteristic impedance of the feeder to give a VSWR not exceeding 1.05:1 at the carrier frequency. At the edge of the modulation band, changes in the antenna impedance and the reactances of the matching network components result in an increase of VSWR. However, the energy at the edges of the modulation band is small and, at these frequencies, a VSWR up to about 1.5:1 can be accepted.

Generally, a T network of three reactances is used to match the impedance of the antenna to the characteristic impedance of the feeder and thus to provide the correct resistive load for the transmitter. A convenient design principle is to cancel the antenna reactance with an equal and opposite compensating reactance and then to transform the resulting resistance to the characteristic impedance of the feeder with a quarter-wave line as illustrated in Figure 2.4(a). In practice, the quarter-wave line and the compensating reactance are combined and take the form of a three-element, lumped constant T network as illustrated in Figure 2.4(b). As an example, Figure 2.4(c) shows the T matching network required for the following typical case:

Carrier frequency	700 kHz
Audio bandwidth	6 kHz
Antenna impedance at carrier	$19.9 - j77.8\ \Omega$
Antenna impedance at carrier + 6 kHz	$20.4 - j75.0\ \Omega$
Antenna impedance at carrier − 6 kHz	$19.5 - j80.7\ \Omega$
Feeder characteristic impedance	50 Ω

The VSWRs at carrier and the carrier frequency and at the extremes of the modulation band will be:

Carrier frequency	1.00:1
Carrier frequency + 6 kHz	1.21:1
Carrier frequency − 6 KHz	1.22:1

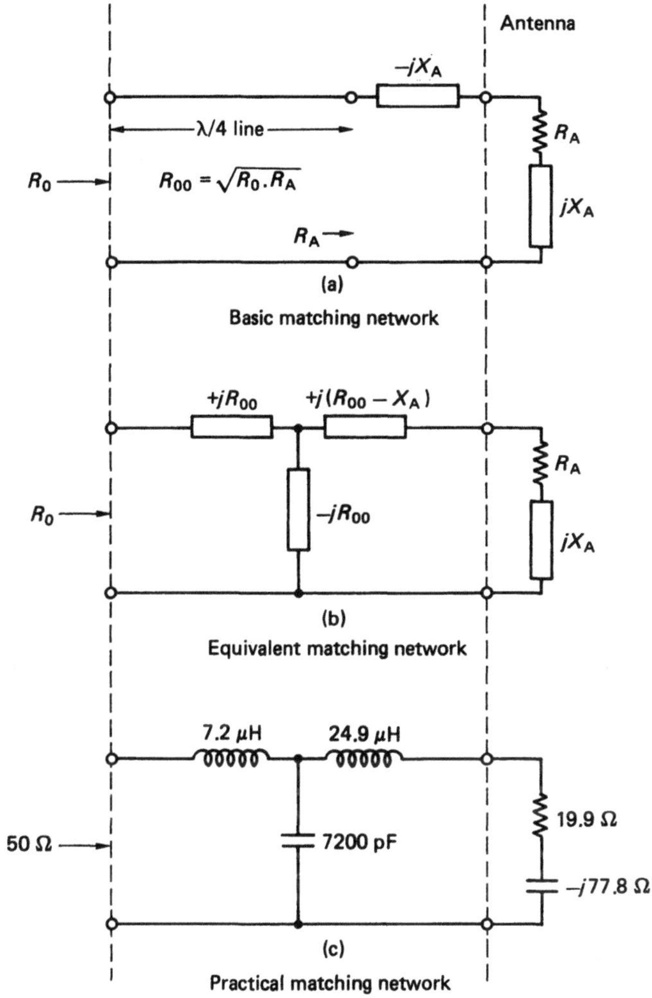

Figure 2.4 Antenna matching circuits

2.3.5 Combining networks

To achieve economy in ground area, and to minimize the number of mast structures required at a transmitting station, it is not uncommon to transmit more than one programme from a single MF antenna. This obviously increases the complexity of the matching arrangements and the values of peak voltages occurring and, for this reason, it is unusual to radiate more than three programmes in this way. To illustrate the design approach, Figure 2.5 shows the matching and combining network arrangement for two transmissions radiated

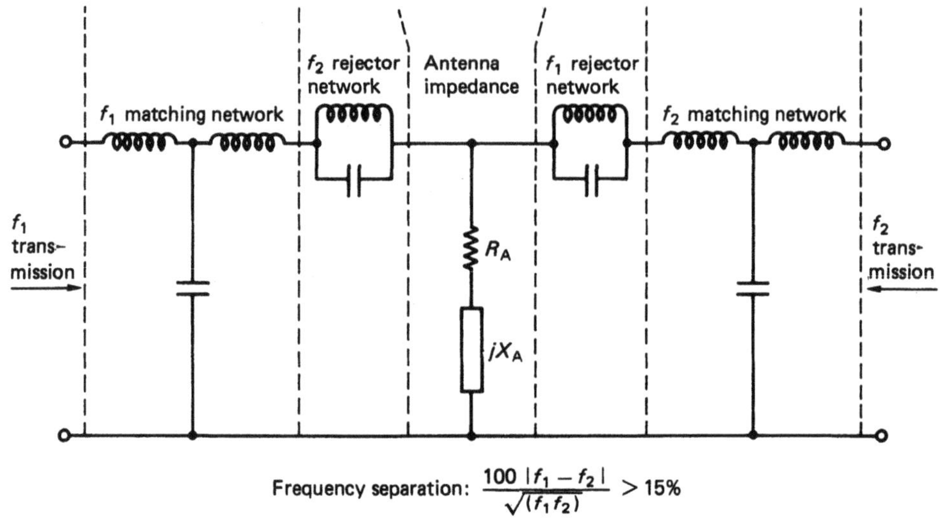

Figure 2.5 Combining network for two MF transmitters. f_1, frequency of left-hand transmitter; f_2, frequency of right-hand transmitter

from the same antenna. It will be seen that a parallel rejector circuit is included in series with the path of each transmission to pass that transmission but to block the other. This is essential to prevent one transmitter feeding energy into the other but, at the same time, to allow the required transmission to pass to the antenna with minimum attenuation. In addition to the parallel resonant circuits, there is a T network in the path of each transmission that matches the antenna impedance (including the effect of the parallel resonant circuit) to the characteristic impedance of the feeder. To avoid undue power losses in the parallel resonant circuits, the carrier frequencies of the two transmissions must be separated by at least 15% as shown in Figure 2.5.

2.3.6 Radiation patterns

In some cases it is necessary to provide a directional radiation pattern in which the transmitted signal varies with azimuth angle. This can occur when the transmitter is not sited in the centre of the area to be served or when it is required to reduce the radiated signal in the direction of stations operating on the same frequency and thus reduce co-channel interference. In these cases, a directional antenna is required which can be achieved by using two or more monopoles driven with appropriate radiating currents. The increase in gain of a directional antenna, in the direction of maximum radiation, over that of a single monopole, depends on the number of monopoles, their height, spacing and method of feeding but falls into the following ranges:

1 monopole (reference)	0 dB
2 monopoles	2–4 dB
3 monopoles	4–6 dB
4 monopoles	5–7 dB
5 monopoles	6–8 dB
6 monopoles	6.5–8.5 dB

The number of independent nulls in the radiation pattern that can be provided to protect co-channel stations is, theoretically, $N-1$, where N is the number of monopoles (for example, a four-monopole antenna would provide three independent nulls). In a directional antenna only one monopole may be driven and the remainder energized parasitically (i.e. carrying currents induced by coupling from the driven monopole). Alternatively, all the monopoles may be driven and this arrangement, although more complex and expensive, gives somewhat higher gain and allows deep nulls to be achieved and located accurately in azimuth. Figure 2.6(a) shows the arrangement and radiation pattern of a typical two-monopole antenna with parasitic reflector, and Figure 2.6(b) gives the corresponding information for a typical two-monopole antenna in which both the monopoles are driven. It will be seen that the radiation patterns are similar but that for the 'driven' antenna has a deeper null.

In some cases there are natural or man-made obstructions in the neighbourhood of MF transmitting sites which may affect the transmission characteristics of the antenna. The presence of an obstruction can modify both the impedance and the radiation pattern of an antenna but, normally, the latter poses the more severe problem. This is because a spacing between the antenna and the obstruction that will reduce distortion of the radiation pattern to an acceptable level will almost certainly have little effect on the impedance. The acceptable level of perturbation to the radiation pattern of an MF antenna will vary with circumstances but, in most cases, will be in the range ± 0.5 dB to ± 1 dB. The effect of an obstruction on the radiation pattern is a function of its projected area (as seen from the antenna), the spacing and the frequency. Typical values of minimum spacing for a frequency of 1 mHz and a perturbation of ± 0.5 dB are shown in Table 2.7.

Table 2.7

Projected area of obstruction (m^2)*	*Required spacing from antenna (m)*
100	5
400	12
900	27
1600	48
2500	80

*The apparent area of the obstruction when viewed from the base of the transmitting mast.

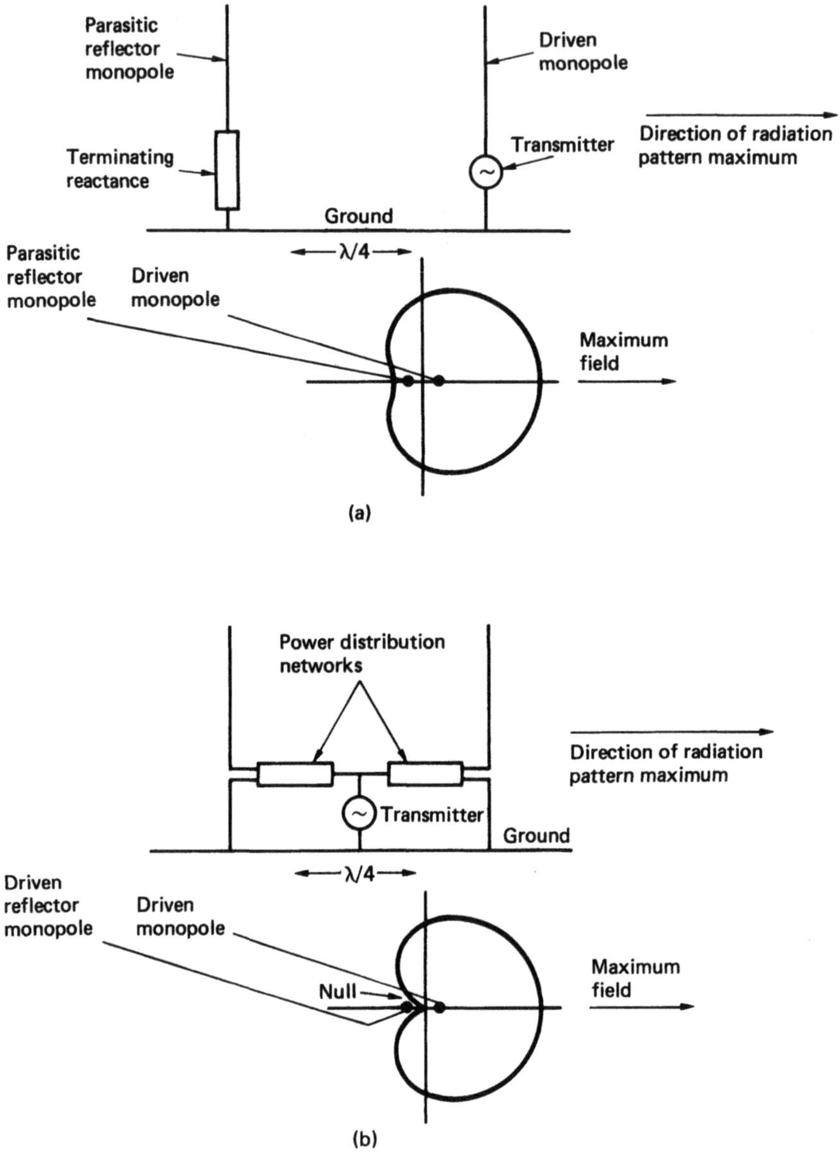

Figure 2.6 Directional MF antennas. (a) Parasitic two-monopole antenna and typical radiation pattern in horizontal plane. (b) Antenna with both monopoles driven and typical radiation pattern in horizontal plane

2.3.7 Reliability

LF and MF antennas are relatively simple both electrically and mechanically and, in general, are highly reliable. Despite this, however, a failure cannot be ruled out. A base insulator, one of the stay insulators or one of the matching components may become faulty and, in any event, an antenna will need to be withdrawn from service occasionally for routine maintenance (e.g. mast or tower painting and the adjustment of stay tensions). Further, although very unlikely, a catastrophic failure due, for example, to freak weather conditions or collision by an aircraft cannot be completely ruled out. In view of the high reliability of MF antennas it is not normal to provide a permanent 'standby' antenna. However, if it is necessary to provide a temporary antenna to maintain the service when the main antenna is not available the following arrangements are possible:

1. For a directional antenna, comprising a number of masts, switching can be provided so that a single mast can be used in the event of a failure of one of the masts.
2. On a site from which a number of transmissions are made, a standby antenna can be provided by attaching a 'sloping wire' to a mast or tower forming part of the antenna of another service.
3. Where there are a number of stations to be considered, it can be economic to provide a transportable antenna system comprising a lightweight mast radiator and a prefabricated ground mat, capable of rapid installation. Such a system can be held available, centrally, for use at any station where a fault has occurred.

It should be realized that none of the above arrangements will provide a standby antenna equal in performance to the main antenna it replaces. But, since radiation of a service from a standby antenna will be a relatively rare event, some deterioration in coverage can be accepted.

Chapter 3

HF propagation and antennas

3.1 Introduction

In the early years of radio transmission it was thought that, however great the radiated power, the earth's curvature would limit the range to a few hundred kilometres. Radio waves, like light waves, travel in straight lines and it seemed obvious that they could not be received round the curve of the earth except by diffraction with consequent very high attenuation. However, in the 1920s it was found that radio transmissions in the HF band (2–30 MHz) could, in fact, be received over distances of several thousand kilometres and that this was due to the presence of layers of ionized gas above the surface of the earth. These layers comprise the ionosphere, which is formed by the radiated energy of the sun, and refracts radio signals in the HF band back to earth, thus making long-distance broadcasting possible.

Once the practicability of long-distance HF broadcasting had been established, the development of transmitting equipment and transmitting antennas commenced. Also, and of equal importance, research was undertaken to establish the characteristics of the ionosphere and the operating frequencies and power required to provide a service over a given path. So far as transmitting antennas are concerned it was, of course, clear that the radiation pattern required for long-distance HF transmission would be quite different from that giving the wide angle, or omnidirectional, coverage required for short-distance domestic broadcasting. The object of HF broadcasting is, normally, to serve a distant area and the signal therefore only needs to be radiated over a relatively narrow range of angles. The radiated energy can therefore be concentrated in a 'beam' with a consequent maximization of signal in the target service area.

3.2 Propagation

3.2.1 Ionospheric propagation fundamentals

Figure 3.1(a) shows that, due to the curvature of the earth, there is a limit to the communication distance that can be achieved with one ionospheric reflection (known as the 'one-hop' mode). However, greater distances can be covered if the signal is first reflected from the ionosphere, then re-reflected from the earth and

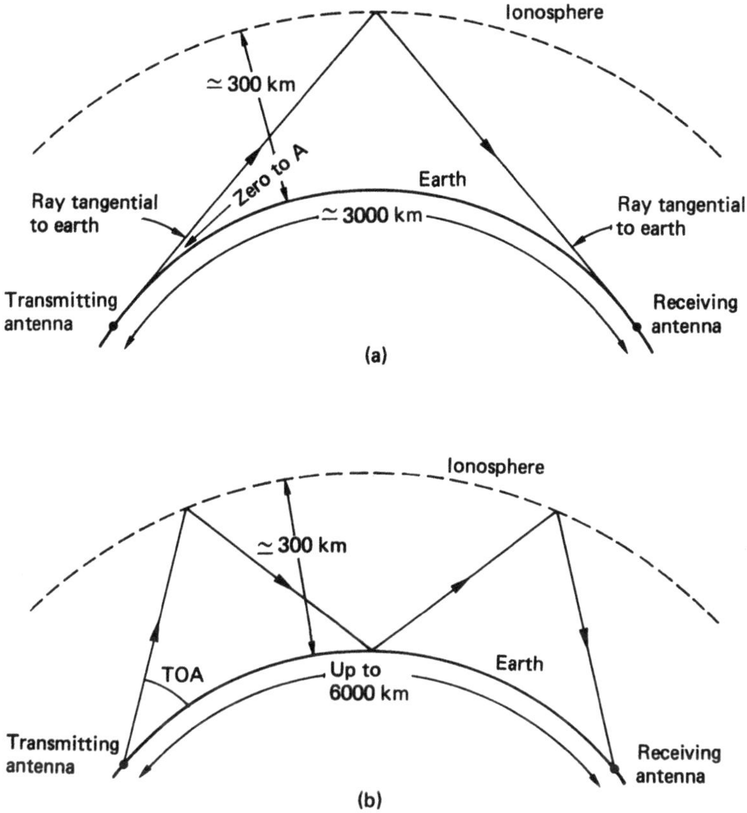

Figure 3.1 (a) Maximum distance of one-hop mode. (b) Transmission by two-hop mode

then re-reflected a second time from the ionosphere (the 'two-hop' mode illustrated in Figure 3.1(b)). Obviously, greater numbers of ionospheric reflections are possible ('three-hop', 'four-hop' modes etc.) and the maximum length of each 'hop' corresponds to a distance of about 3000 km. In practice, the signal quality over paths with more than two hops (i.e. distances exceeding 6000 km) tends to be unreliable but broadcasting over four-hop paths (distances around 12 000 km) is not unknown. It is worth pointing out, however, that over very long distances the transmission becomes diffuse and a division of the signal path into separate hops tends to break down.

The ionosphere is not a simple reflector and comprises four distinct layers designated D, E, F1 and F2 with approximate average heights, respectively, of 100 km, 140 km, 200 km and 350 km. The mechanism whereby radio waves are returned to earth is actually one of total internal refraction but, for simplicity, it is commonly referred to as reflection from the effective height (i.e. the apparent height) of the ionospheric layer concerned. During the daytime the D-layer absorbs radio waves with frequencies below about 2 MHz and this is why HF broadcasting uses frequencies above 2 MHz. The E-layer reflects MF transmission at night but plays little part in HF broadcast transmission. During the daytime, HF transmissions are predominantly via the F2-layer but, during the night-time, the F1- and F2-layers combine to form a single F-layer which is the predominant HF-reflecting surface. The heights of the daytime F2-layer and night-time F-layer vary between 300 km and 500 km, depending on latitude and time of year.

The layers of the ionosphere are formed by ultraviolet radiation from the sun and, as might be expected, the ionization varies in intensity with latitude, longitude, time of day (diurnal change) and time of year (seasonal change). Ionization has also been found to vary over the 11 year cycle of sunspot activity and can be predicted using the 'sunspot number', which has been found to vary from a low of around 5 to around 160. The sunspot number is not the number of sunspots, but a number derived by comparing the total area of sunspots averaged over a period of time – usually a month – with the area of the sun, and past records and forecasts are published by the ITU (International Telecommunications Union).

From time to time solar flares occur, apparently originating in sunspots, and give rise to intense ionization and magnetic storms. Under these conditions transmissions of the lower frequencies are greatly attenuated, as are transmissions over polar regions, and the aurora may be visible.

The reflecting properties and effective height of each layer can be determined by 'ionospheric sounding'. This involves transmitting a short pulse at a given frequency (a 'wave packet') vertically and measuring the level and delay of the pulse reflected back to earth from the ionosphere. For each ionospheric layer, and starting at a low frequency, an increase in the frequency of the pulse results in an increase in the level of the reflected signal. This increase continues until a frequency (known as the critical frequency) is reached at which the attenuation reaches a minimum. At frequencies above the critical frequency the attenuation of the reflected pulse increases rapidly due to penetration of the ionosphere. In HF broadcasting, of course, the signal does not arrive at the ionosphere vertically (i.e. with zero angle of incidence) and this affects the frequency at which the minimum attenuation occurs. For non-zero angles of incidence the frequency of minimum attenuation is known as the MUF (maximum usable frequency) and is given approximately by the critical frequency divided by the cosine of the angle of incidence.

The MUF for a given path varies due to short-term ionospheric changes with consequent decreases and increases in attenuation. To allow for these variations a frequency about 15% less than the MUF is used for transmission and is known as the FOT (frequency of optimum transmission). The values of FOT used in HF broadcasting vary from a minimum of about 2 MHz up to a maximum of about 27 MHz (higher frequencies are not reflected effectively from the ionosphere under normal conditions).

For broadcasting over an HF path it is obviously necessary that the radiated signal intercepts the target reception area after reflection from the ionosphere. It is easy to see, therefore, that, for maximum received signal, the energy radiated by the transmitting antenna needs, ideally, to be a maximum at a specific angle to the horizontal which is known as the 'path TOA' (the path take-off angle). The angle to the horizontal at which the antenna radiation is actually a maximum is known as the 'antenna TOA' and, clearly, must match the path TOA as closely as is possible.

It is important to note that during refraction in the ionosphere the polarization of the transmitted signal is modified. As a result, irrespective of the transmitted polarization, the received signal comprises approximately equal vertically polarized (VP) and horizontally polarized (HP) components. It follows, therefore, that the polarization of the transmitting and receiving antennas can be chosen independently on grounds of practical convenience. Most HF transmitting antennas are HP since this obviates the need for an expensive ground screen and results in the maximum radiation efficiency. However, VP antennas are, occasionally, used in cases where a very low antenna TOA or a local ground-wave signal is required.

The argument concerning the need to 'match' the path TOA and antenna TOA at the transmitting point can be applied to the reception of HF signals although, over very long multi-hop paths, the transmitted and received values of path TOA may be different. In the case of communications links where, for reasons of economy, transmitter power is kept to a minimum, every effort is made to match, precisely, the receiving antenna TOA to the path TOA at the receiving point. However, in the case of broadcasting, the receiving antenna will normally be the simple 'whip' of the listener's receiver so that it is necessary to use high transmitted powers to achieve a service. It is important to note that the

Table 3.1

Kilohertz	*Megahertz band*	*Metre band*
2 300–2 495	2	120*
3 200–3 400	3	90*
3 900–4 000	4	75†
4 750–5 060	5	60*
5 950–6 200	6	49
7 100–7 300	7	41†
9 500–9 775	9	31
11 700–11 975	11	25
15 100–15 450	15	19
17 700–17 900	17	16
21 450–21 750	21	13
25 600–26 100	26	11

*For use only in designated tropical zones.
†Not used for broadcasting in the western hemisphere.

signal-to-noise ratio (SNR) in HF reception is due to external noise (as opposed to receiver noise) and is therefore controlled by receiving antenna directivity as opposed to gain.

3.2.2 The HF broadcast frequency bands

The frequency bands used for HF broadcasting are allocated under agreements drawn up under the auspices of the ITU (International Telecommunications Union) in Geneva. Under these agreements the bands shown in Table 3.1 were allocated in 1979 for broadcasting.
The additional bands shown in Table 3.2 are allocated to the HF broadcast band and are used by some broadcasters.

Table 3.2

Kilohertz	*Kilohertz*
9 775–9 900	15 450–15 600
11 650–11 700	17 550–17 700
11 975–12 050	21 750–21 850

The energy received from the sun varies with the time of day, time of year, latitude and sunspot number and the FOT for a given path therefore varies. As an example, the FOTs for a typical 5000-km circuit in the northern hemisphere are shown in Table 3.3.

Table 3.3

	FOT (MHz)	
	At sunspot maximum	*At sunspot minimum*
Winter day	22	13
Winter night	7	5
Summer day	23	18
Summer night	16	9

It will be seen, therefore, that the maintenance of a single long-distance service can require frequencies covering most of the HF broadcast band.

In a country with a large goegraphical area, a very large number of conventional MF or VHF transmitting stations will be needed to provide 'blanket' coverage. However, in such cases coverage can be provided economically by HF 'sky-wave' broadcasting. With this technique, the signal is transmitted vertically, and returned to earth after reflection from the ionosphere. An area extending from the neighbourhood of the station out to distances of 1000 km or more can be covered and, as a result, a single transmitting station can serve an area 100, or more, times greater than a single MF or VHF station. As with all ionospheric transmission, it is necessary for a range of frequencies to be used and the listener must be prepared to switch between channels. Nevertheless, despite this disadvantage, particularly for speech and light entertainment programmes, HF sky-wave broadcasting offers an economic and effective means of providing a service over a very large area. Because sky-wave broadcasting involves transmission that is short range (by normal HF broadcast standards), operating frequencies low in the HF band are required, even at the peak of the sunspot cycle. For this reason, as mentioned above, frequencies in the 2–5-MHz segment of the HF broadcast band are reserved for short- and medium-range broadcasting in designated tropical areas.

3.2.3 HF propagation analysis

As in other frequency bands, propagation analysis in the HF band aims, primarily, at determining received field strength in terms of transmitter power, antenna gain and frequency. However, because of the variation of the characteristics of the ionosphere with frequency and time, the methods of analysis are complex and the predictions can never be precise. The overall measure of effectiveness of communication is given by the 'reliability', which is defined as the percentage of transmission time during which the SNR, at the receiver output, equals or exceeds a required minimum. Thus, for the calculation of reliability, it is necessary to take account of the gain of the transmitting antenna and the directivity of the receiving antenna as a function of azimuth and elevation angle. However, for a preliminary propagation analysis, it is often convenient to assume the use of isotropic transmitting and receiving antennas (a transmitting antenna which radiates equally in all directions or a receiving antenna with equal sensitivity in all directions).

Virtually all HF propagation analysis computer programs are based on a geometrical model comprising an ionosphere and earth which each absorb part of the incident energy and reflect the remainder. A propagation analysis program is based on a field strength calculated from the fundamental formula:

$$FS = 20 \log [(3W/d)^{0.5} \times 10^5] + G - L_x - L_g \, \mathrm{dB}\mu$$

where FS is the field strength in dBμ, W is the transmitter power in kW, d is the path length in km, G is the antenna gain in dBi, L_x is the ionospheric loss in dB, and L_g is the ground loss in dB. In the formula, the variables are known (or are easy to calculate) except for the ionospheric loss, which depends on frequency, path TOA, path mode (number of hops), time of day, time of year, latitude, longitude and sunspot number. Computer programs are available in which data are stored and updated concerning critical frequency, ionospheric height and ionospheric loss on a worldwide basis over the daytime, seasonal and sunspot cycles. Such programs can only be operated on fairly large high-speed computers and, even then, the predictions can only be regarded as approximate. Useful predictions of FOT, field strength and reliability can, however, be provided by simpler programs that can be operated on a PC and the example that follows has been found in practice to give results which are in reasonable agreement with those of more complex programs. The program used in the example is designed to include, if required, the gain and directivity of the transmitting and receiving antennas as functions of the azimuth and elevation angles.

Figures 3.2(a) and (b) show printouts of a typical propagation analysis program for broadcasting over a 3000-mile (5172-km) path in the northern hemisphere. The two printouts represent the extremes of the seasonal and sunspot conditions which are taken as SSN 120 for June and SSN 20 for December. The detailed circuit parameters are described at the top of each page and the first line gives the program version (in this case HFFOT/2).

The second line contains the transmitting and receiving site names and the great circle path distance in kilometres. The third line gives the sunspot number

```
HFFOT/2 (HF PROPAGATION ANALYSIS)
PATH: FROM DAVENTRY TO BOSTON    DISTANCE 5172.00 km
SSN 120.00    MONTH june    MID-PATH LAT 55.00 deg MID-PATH LONG 325.00 deg
TRANSMIT ANTENNA: GAIN 22.00 dbi TYPE HR 4/4/i
RECEIVE ANTENNA: DIRECTIVITY 0.00 dbi    TYPE: constant-gain
REQUIRED SNR 75.00 DB    RECEIVE NOISE LEVEL suburban
TRANSMITTER POWER 100.00 KW

AVERAGE PATH TOA 9.29 DEG
PRINCIPAL MODE 2.00F

    0.00 hours gmt
FOT = 15.18 MHZ
   15.18    2.00    4.00    8.00   12.00   16.00   25.00   30.00   MHZ
   56.88   50.28   52.54   54.79   56.11   55.16   40.62   34.68   DBU
   89.77   67.34   77.37   84.89   88.19   71.64   10.65    4.89   REL%

    2.00 hours gmt
FOT = 14.45 MHZ
   14.45    2.00    4.00    8.00   12.00   16.06   25.00   30.00   MHZ
   56.88   50.44   52.70   54.95   56.27   53.56   39.03   33.09   DBU
   89.77   68.13   77.99   85.33   88.54   58.12    8.64    3.97   REL%

    4.00 hours gmt
FOT  14.42 MHZ
   14.42    2.00    4.00    8.00   12.00   16.00   25.00   30.00   MHZ
   56.88   50.44   52.70   54.96   56.28   53.49   38.96   33.02   DBU
   89.77   68.16   78.01   85.35   88.55   57.60    8.57    3.93   REL%

    6.00 hours gmt
FOT = 15.09 MHZ
   15.09    2.00    4.00    8.00   12.00   16.00   25.00   30.00   MHZ
   56.88   50.30   52.55   54.81   56.13   54.98   40.44   34.50   DBU
   89.77   67.43   77.44   84.94   88.23   69.97   10.41    4.78   REL%

    8.00 hours gmt
FOT = 16.29 MHZ
   16.29    2.00    4.00    8.00   12.00   16.00   25.00   30.00   MHZ
   56.88   50.05   52.31   54.56   55.88   56.82   42.93   36.99   DBU
   89.77   66.19   76.46   84.24   87.66   89.66   14.41    6.62   REL%
```

10.00 hours gmt
FOT = 17.69 MHZ

17.69	2.00	4.00	8.00	12.00	16.00	25.00	30.00	MHZ
56.88	49.78	52.04	54.30	55.62	56.55	45.61	39.67	DBU
89.77	64.82	75.37	83.44	87.03	89.12	20.48	9.40	REL%

12.00 hours gmt
FOT = 18.91 MHZ

18.91	2.00	4.00	8.00	12.00	16.00	25.00	30.00	MHZ
56.88	49.56	51.82	54.08	55.40	56.33	47.79	41.85	DBU
89.77	63.69	74.45	82.77	86.49	88.67	27.27	12.52	REL%

14.00 hours gmt
FOT = 19.64 MHZ

19.64	2.00	4.00	8.00	12.00	16.00	25.00	30.00	MHZ
56.88	49.44	51.70	53.95	55.27	56.21	49.02	43.08	DBU
89.77	63.04	73.93	82.39	86.18	88.40	32.02	14.70	REL%

16.00 hours gmt
FOT = 19.67 MHZ

19.67	2.00	4.00	8.00	12.00	16.00	25.00	30.00	MHZ
56.88	49.43	51.69	53.95	55.27	56.21	49.07	43.13	DBU
89.77	63.02	73.91	82.37	86.17	88.39	32.24	14.80	REL%

18.00 hours gmt
FOT = 19.00 MHZ

19.00	2.00	4.00	8.00	12.00	16.00	25.00	30.00	MHZ
56.88	49.55	51.80	54.06	55.38	56.32	47.93	42.00	DBU
89.77	63.61	74.39	82.73	86.45	88.63	27.79	12.76	REL%

20.00 hours gmt
FOT = 17.80 MHZ

17.80	2.00	4.00	8.00	12.00	16.00	25.00	30.00	MHZ
56.88	49.76	52.02	54.27	55.59	56.53	45.82	39.88	DBU
89.77	64.71	75.28	83.38	86.98	89.08	21.05	9.66	REL%

22.00 hours gmt
FOT = 16.40 MHZ

16.40	2.00	4.00	8.00	12.00	16.00	25.00	30.00	MHZ
56.88	50.03	52.28	54.54	55.86	56.80	43.15	37.21	DBU
89.77	66.08	76.37	84.17	87.61	89.61	14.85	6.82	REL%

Figure 3.2 Propagation analysis. (a) High sunspot number.

HFFOT/2 (HF PROPAGATION ANALYSIS)

PATH: FROM DAVENTRY TO BOSTON DISTANCE 5172.00 km
SSN 20.00 MONTH dec MID-PATH LAT 55.00 deg MID-PATH LONG 325.00 deg
TRANSMIT ANTENNA: GAIN 22.00 dbi TYPE HR 4/4/1
RECEIVE ANTENNA: DIRECTIVITY 0.00 dbi TYPE: constant-gain
REQUIRED SNR 75.00 DB RECEIVE NOISE LEVEL suburban
TRANSMITTER POWER 100.00 KW

AVERAGE PATH TOA 6.36 DEG
PRINCIPAL MODE 2.00F

0.00 hours gmt
FOT = 7.66 MHZ

7.66	2.00	4.00	8.00	12.00	16.00	25.00	30.00	MHZ
55.60	51.23	53.48	54.18	40.97	31.60	17.07	11.13	DBU
86.99	71.85	80.84	72.21	12.79	3.74	0.56	0.26	REL%

2.00 hours gmt
FOT = 6.14 MHZ

6.14	2.00	4.00	8.00	12.00	16.00	25.00	30.00	MHZ
55.60	51.95	54.20	46.97	33.77	24.40	9.86	3.92	DBU
86.99	74.99	83.17	28.08	4.97	1.46	0.22	0.10	REL%

4.00 hours gmt
FOT = 6.07 MHZ

6.07	2.00	4.00	8.00	12.00	16.00	25.00	30.00	MHZ
55.60	51.98	54.24	46.63	33.43	24.06	9.52	3.58	DBU
86.99	75.14	83.27	26.85	4.75	1.39	0.21	0.10	REL%

6.00 hours gmt
FOT = 7.48 MHZ

7.48	2.00	4.00	8.00	12.00	16.00	25.00	30.00	MHZ
55.60	51.30	53.56	53.43	40.22	30.85	16.31	10.38	DBU
86.99	72.19	81.09	65.44	11.59	3.39	0.50	0.23	REL%

8.00 hours gmt
FOT = 9.99 MHZ

9.99	2.00	4.00	8.00	12.00	16.00	25.00	30.00	MHZ
55.60	50.36	52.62	54.88	49.62	40.25	25.72	19.78	DBU
86.99	67.75	77.70	85.12	39.74	11.63	1.73	0.79	REL%

10.00 hours gmt
FOT = 12.92 MHZ

12.92	2.00	4.00	8.00	12.00	16.00	25.00	30.00	MHZ
55.60	49.52	51.78	54.04	55.36	48.63	34.09	28.15	DBU
86.99	63.49	74.29	82.66	86.40	34.88	5.19	2.38	REL%

12.00 hours gmt
FOT = 15.49 MHZ

15.49	2.00	4.00	8.00	12.00	16.00	25.00	30.00	MHZ
55.60	48.93	51.19	53.45	54.77	54.54	40.00	34.06	DBU
86.99	60.34	71.69	80.72	84.82	75.67	11.25	5.17	REL%

14.00 hours gmt
FOT = 17.01 MHZ

17.01	2.00	4.00	8.00	12.00	16.00	25.00	30.00	MHZ
55.60	48.63	50.89	53.14	54.46	55.40	43.05	37.11	DBU
86.99	58.67	70.28	79.65	83.95	86.50	16.78	7.71	REL%

16.00 hours gmt
FOT = 17.07 MHZ

17.07	2.00	4.00	8.00	12.00	16.00	25.00	30.00	MHZ
55.60	48.62	50.87	53.13	54.45	55.39	43.17	37.23	DBU
86.99	58.60	70.22	79.60	83.91	86.47	17.06	7.83	REL%

18.00 hours gmt
FOT = 15.66 MHZ

15.66	2.00	4.00	8.00	12.00	16.00	25.00	30.00	MHZ
55.60	48.90	51.15	53.41	54.73	54.90	40.36	34.43	DBU
86.99	60.14	71.52	80.59	84.72	79.38	11.81	5.42	REL%

20.00 hours gmt
FOT = 13.16 MHZ

13.16	2.00	4.00	8.00	12.00	16.00	25.00	30.00	MHZ
55.60	49.46	51.72	53.98	55.30	49.23	34.69	28.75	DBU
86.99	63.18	74.04	82.47	86.24	37.72	5.61	2.58	REL%

22.00 hours gmt
FOT = 10.23 MHZ

10.23	2.00	4.00	8.00	12.00	16.00	25.00	30.00	MHZ
55.60	50.28	52.54	54.80	50.39	41.02	26.49	20.55	DBU
86.99	67.37	77.40	84.91	43.96	12.87	1.91	0.88	REL%

(b) Low sunspot number

(SSN), the month and the mid-path latitude and mid-path longitude. The fourth line gives the transmitting antenna gain in dBi and the type (this can be a maker's designation or constant gain). The fifth line gives the receiving antenna directivity in dBi and the type (maker's designation or constant gain). The sixth line gives the required SNR (which is 75 dB in a 1-Hz bandwidth for a first class service and 71 dB for a second class service) and the receive noise level (rural, suburban or urban). The seventh line gives the transmitter power (carrier power for DBS or PEP for SSB).

The eighth line gives the calculated path TOA averaged over the 24-hour period and the ninth line gives the principal mode (1.00F means one-hop via the F(F1) layer, 2.00F means two hops etc.).

The first line of each time-block in the body of the table gives the universal time (GMT) and the second line gives the FOT. The third line is a list of the frequencies used in the analysis and these can be adjusted to any required values. The fourth and fifth lines give the field strengths and reliabilities corresponding to the frequencies given in the third line.

The program gives a full propagation analysis in terms of the 24-hour cycle and for any combination of sunspot number and month of the year. However, each printout concerns only one path length and two printouts are needed for analysis at the extremes of the sunspot cycle. In many cases, however, an analysis is required for a range of receiving points and, for this case, the full program is laborious and time-consuming. A separate program has therefore been written to provide a rapid summary propagation analysis for a range of distances from a transmitting site. An example of a propagation analysis printout from this program, for a high-grade programme link from Kuwait, is shown in Figure 3.3.

```
HFFOT/3 - HF PROPAGATION SUMMARY

FROM: KUWAIT     TO: VARYING DISTANCES
SYSTEM SNR: 70.00 dB    RECEIVE NOISE LEVEL: SUBURBAN
TRANSMITTER POWER: 10.00 KW
TRANSMIT ANTENNA GAIN: 15.00 DBI    RECEIVE ANTENNA DIRECTIVITY: 5.00 DBI

                    DISTANCE: 500.00 KM

SSN: 120.00    MONTH: JUNE
AVERAGE TOA: 52.20 DEG    MODE: 1.00F    FOTMAX; 11.59 MHz    FOTMIN 6.93 MHz
AVERAGE FIELD-STRENGTH: 60.55 DBU        RELIABILITY: 98.56%

SSN: 20.00    MONTH: DECEMBER
AVERAGE TOA: 47.94 DEG    MODE: 1.00F    FOTMAX: 9.36 MHz    FOTMIN 3.62 MHz
AVERAGE FIELD-STRENGTH: 61.39 DBU        RELIABILITY; 98.64%

                    DISTANCE: 1000.00 KM

SSN: 120.00    MONTH: JUNE
AVERAGE TOA: 31.62 DEG    MODE: 1.00F    FOTMAX: 14.81 MHz    FOTMIN 8.86 MHz
AVERAGE FIELD-STRENGTH: 57.39 DBU        RELIABILITY: 98.10%

SSN: 20.00    MONTH: DECEMBER
AVERAGE TOA: 27.70 DEG    MODE: 1.00F    FOTMAX: 12.28 MHz    FOTMIN 4.75 MHz
AVERAGE FIELD-STRENGTH: 57.80 DBO        RELIABILITY: 98.18%

                    DISTANCE: 1500.00 KM

SSN: 120.00    MONTH: JUNE
AVERAGE TOA: 20.71 DEG    MODE: 1.00F    FOTMAX: 18.15 MHz    FOTMIN 10.85 MHz
AVERAGE FIELD-STRENGTH: 54.70 DBU        RELIABILITY: 97.35%

SSN: 20.00    MONTH: DECEMBER
AVERAGE TOA: 17.62 DEG    MODE: 1.00F    FOTMAX: 15.13 MHz    FOTMIN 5.85 MHz
AVERAGE FIELD-STRENGTH: 54.93 DBU        RELIABILITY: 97.43%

                    DISTANCE: 2000.00 KM

SSN:120.00  MONTH:JUNE
AVERAGE TOA: 14.01 DEG    MODE: 1.00F    FOTMAX: 21.38 MHz    FOTMIN 12.78 MHz
AVERAGE FIELD-STRENGTH: 52.53 DBU        RELIABILITY: 96.31%

SSN: 20.00    MONTH: DECEMBER
AVERAGE TOA: 11.53 DEG    MODE: 1.00F    FOTMAX: 17.82 MHz    FOTMIN 6.88 MHz
AVERAGE FIELD-STRENGTH: 52.68 DBU        RELIABILITY: 96.40%
```

Figure 3.3 Propagation analysis summary

3.3 Antennas

3.3.1 The half-wave dipole and multi-element antennas

In the majority of applications an HF antenna is required to radiate a 'beam' of energy in the direction of the target service area in order to provide maximum signal strength. A beam can be achieved with an HF antenna comprising a number of basic antenna elements (usually half-wave dipoles) disposed physically, and driven with currents of such amplitudes and phases that their radiated signals add in the desired direction and tend to cancel in other directions. There are a number of configurations in which half-wave dipoles can be arranged to generate a beam and the principal methods are the 'curtain antenna' and the 'log-periodic antenna' (the LPA). Both these antennas make use of signal reflected from the earth plane.

3.3.2 Polarization and the earth plane

It is well known that, in the absence of the earth, a horizontal dipole, viewed end-on, has a circular radiation pattern in the vertical plane. However, when mounted above the earth, the VRP is modified so that a number of lobes (often called 'grating lobes') are formed as shown in Figure 3.4(a) for the case of a dipole at a typical height of 1.5 wavelengths. If a number of dipoles are stacked vertically, with the same mean height as the single dipole, the lobes, above the lowest, are reduced in amplitude. Thus, if four dipoles are stacked vertically, as shown in Figure 3.4(b), a main lobe is formed at an elevation angle known as the 'take-off angle' (TOA). The enhancement of the lowest grating lobe and the consequent formation of a main lobe can also be achieved if a number of dipoles are arranged in 'end-fire' as shown in Figure 3.4(c). It has been pointed out that an HF transmitting antenna needs to radiate maximum power at a specific TOA so that, after reflection by the ionosphere, the signal intercepts the target service area. It can be shown that the TOA of an antenna increases as the mean height is decreased and decreases as the mean height is increased and this makes it possible to achieve a TOA suited to a particular path length. For example, a mean height of 1.5 wavelengths, as shown in Figure 3.4(a), will provide a TOA of 8°, suitable for single-hop transmission over path lengths in the region of 2500 km.

3.3.3 HF curtain antennas

The great majority of HF broadcast transmitting antennas are based on a rectangular curtain of HP (horizontally polarized) half-wave dipoles driven with

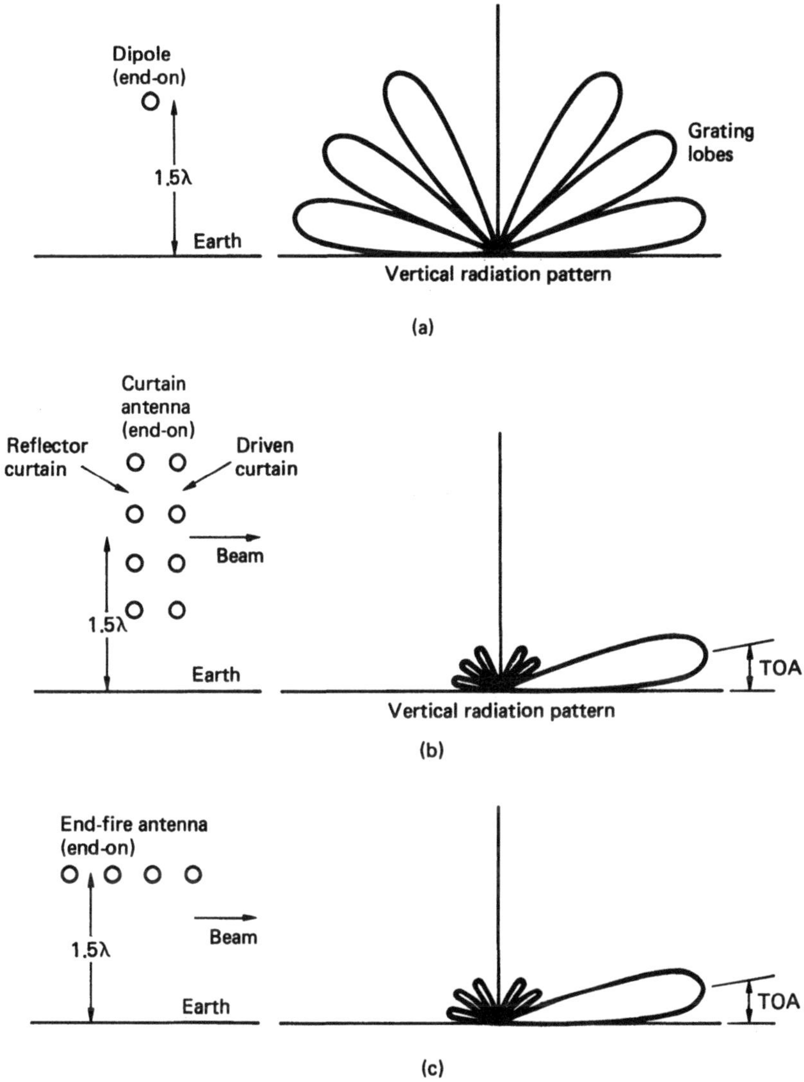

Figure 3.4 Radiation patterns of antennas above earth. (a) Dipole. (b) Curtain antenna. (c) Progressively phased end-fire antenna

equal-amplitude, co-phased currents as illustrated in Figure 3.5. Horizontal (as opposed to vertical) dipoles are used as they have the advantage that the earth losses are negligible. It is possible to use vertical dipoles but, even if a large and expensive earth screen is provided, there will still be a significant loss of received signal.

A single curtain will radiate in both directions so that a reflector curtain must be used to achieve a unidirectional (single-beam) radiation pattern. A basic

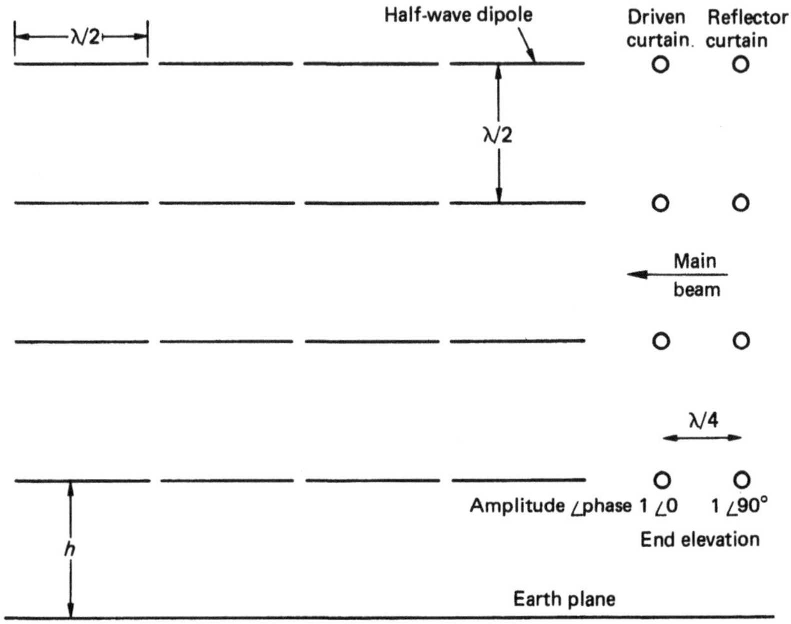

Figure 3.5 Basic curtain antenna

broadcast curtain antenna therefore comprises two curtains, one being driven with equal, co-phased currents in all the dipoles and the other being a parasitic reflector as illustrated in Figure 3.5. The parasitic reflector curtain suppresses the unwanted beam in the direction opposite to that of the service area and increases the gain in the direction of the service area.

The angular width in the horizontal plane of the main lobe of a curtain antenna is inversely proportional to the horizontal aperture. The angular width of the main lobe in elevation, and the TOA, depend on the vertical aperture and the height above ground. Thus by appropriate choice of the horizontal aperture (number of dipoles in a row), vertical aperture (number of rows of dipoles stacked vertically) and height above ground it is possible to control, independently, the horizontal and vertical radiation patterns on a bearing normal to the plane of the curtain. In addition, within limits, it is possible to 'slew' (i.e. varying in bearing) the direction of maximum radiation by linearly varying the phase of the currents fed to the dipoles across the width of the curtain. As a result, the curtain antenna provides a higher degree of operational flexibility than any other type of HF broadcast antenna.

Curtain antennas are classified by the internationally accepted code HRRS/m/n/h where in order the symbols denote:

H – horizontal polarization.
R – a reflector curtain.
R – that the functions of the two curtains can be interchanged so that the beam direction can be reversed.

S – that the beam can be slewed electrically in azimuth.
m – the number of dipoles in a horizontal row.
n – the number of rows stacked vertically.
h – the height of the bottom row of dipoles in wavelengths.

The second R (reversibility) and the S (slewing) are omitted when they are not applicable. Thus the antenna of Figure 3.5 is type HR 4/4/h.

The gain of a curtain antenna depends, principally, on the number of dipoles used in the driven curtain ranges from around 16.5 dBi for an HRS 2/2/h (4 dipoles), to 21 dBi for an HRS 4/4/h (16 dipoles) and 26 dBi for an HRS 6/6/h (36 dipoles).

The bandwidth and azimuth slew of a curtain antenna are limited by changes in gain, radiation pattern, input impedance (VSWR) and TOA as the frequency departs from the design value. These parameters are interdependent and it was not until the 1970s that advances in computer modelling made it possible to carry out the complex calculations necessary to develop wideband, wide-slew, curtain antenna designs. Using computer programs that solve the fundamental electromagnetic equations, and facilitate the structural design process, curtain antennas have now been developed with wider bandwidths, greater slew angles and higher power-carrying capacity than were possible with earlier design techniques. An example of these wideband curtain antennas, developed by Technology for Communications International (TCI) using their electromagnetic computer program (WIRA) and structural computer program (STRUC), is the model 611 illustrated in Figure 3.6. A branch distribution feeder system is used so that each half-wave dipole is fed individually via a path that is identical as regards electrical length and impedance. The dipoles are of the 'folded' type and give a nearly constant resistive impedance over a wide frequency band due to the use of multi-wire radiators and internal reactance compensation. In Figure 3.5 the reflector curtain is shown as similar to the driven curtain and the required phase angle is obtained by a reactive termination. However, this type of reflector curtain becomes rapidly de-tuned as the frequency departs from the design value with a consequent loss of gain and reduction of front-to-back ratio. The wideband curtain antenna therefore uses an 'aperiodic' reflector curtain comprising close-spaced horizontal wires and this enables the gain and front-to-back ratio to be maintained over a very wide frequency band. With this type of reflector curtain the direction of the main lobe cannot be reversed but bi-directional operation can be achieved by erecting two driven curtains (one on either side of the reflector curtain).

The TCI wideband model 611 curtain antenna has a bandwidth of about one octave, which is sufficient to cover four or five adjacent HF broadcasting bands. The HF broadcasting bands can therefore be covered by only two antennas: one for the 6, 7, 9 and 11 MHz bands and the other for the 13, 15, 17, 21 and 26 MHz bands. Further, the fact that the antenna is slewed by progressive phase shift of the four columns, as shown in Figure 3.6, enables nominal slew angles of up to $\pm 30^\circ$ to be obtained without the generation of unacceptably high subsidiary lobes. The nominal driving point impedance of each of the folded dipoles is

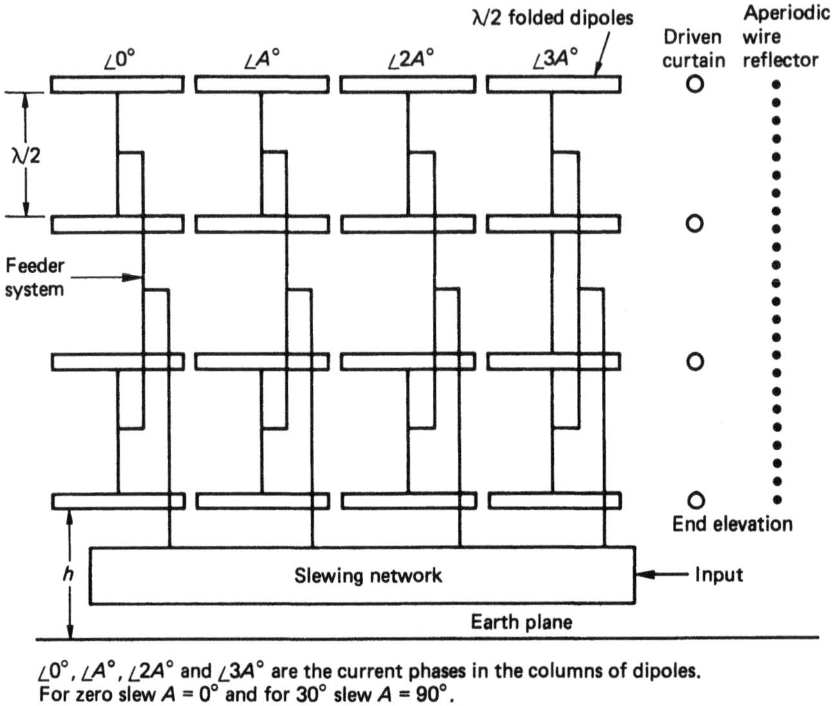

Figure 3.6 Wideband wide-slew curtain antenna

300 Ω and the branch feeder system incorporates wideband, balanced impedance transformers to provide a nominal antenna input impedance of 300 Ω.

As the number of dipoles in a curtain antenna is increased, the mast heights and the complexity of the branch feeder system increase. For this reason the driven curtain of most antennas currently in service does not exceed 16 half-wave dipoles arranged in the HRS 4/4/h configuration illustrated in Figure 3.6. This type of antenna is in extensive use for long-distance HF broadcasting and *h* (the height above ground) ranges from 0.5 to 1.0 wavelengths (i.e. the antennas range from type HRS 4/4/0.5 to type HRS 4/4/1). These antennas, driven by a 500-kW transmitter, will give HF coverage over path distances ranging from 1500 km to 6000 km and above depending on the height.

It has been pointed out that, to serve an area at a particular distance, the antenna must radiate energy at a specific angle to the horizontal (the TOA). This means that if services are to be provided to target areas at widely differing distances, the TOA must be capable of adjustment (i.e. the antenna must be slewable in the vertical plane). Just as horizontal slewing enables target areas over a wide range of azimuth angles to be served, so vertical slewing will enable areas over a wide range of distances to be served by a single antenna.

It is possible to slew the beam of a curtain antenna in the vertical plane by disconnecting one or more of the rows of dipoles or, alternatively, by reversing the phase of one or more rows. In the former case the mean electrical height is changed, and in the latter case the radiation pattern of the antenna (neglecting the effect of the ground plane) is adjusted to favour not the lowest but one of the higher lobes shown in Figure 3.4(a). This principle has been developed by TCI in the design of their model 611vs vertically slewed curtain antenna, which is assembled from a number of modules, each of which is a self-contained HRS 2/2 curtain rather than a single dipole. This is because a smaller module (i.e. one comprising less than four dipoles) will not achieve an acceptable VSWR over an octave bandwidth. The antenna is available in either an HRS 4/4/h or HRS 4/6/h configuration and the height above ground (normally in the range of 0.5–1.0 wavelengths) can be chosen to meet particular operational requirements. The HRS 4/6/h configuration, although more complex, has the advantage that it provides a TOA ranging from over 30° (required for short path lengths around 1000 km) down to below 5° (required for long-distance broadcasting to areas at distances of 6000 km and above from the transmitting station).

The HRS 4/4/h and HRS 4/6/h vertically slewed antennas are also slewable in azimuth by a progressive phase shift of the four columns of dipoles and up to 15 slew angles with a nominal maximum of ±30° are available. Both the vertical and azimuth slew switches are motorized and can be remotely controlled.

3.3.4 Log-periodic wideband antennas

The LPA comprises a series of dipoles of increasing length spaced along a transposed transmission line as illustrated in Figure 3.7. At any given frequency, only dipoles whose lengths are approximately a half-wavelength long are

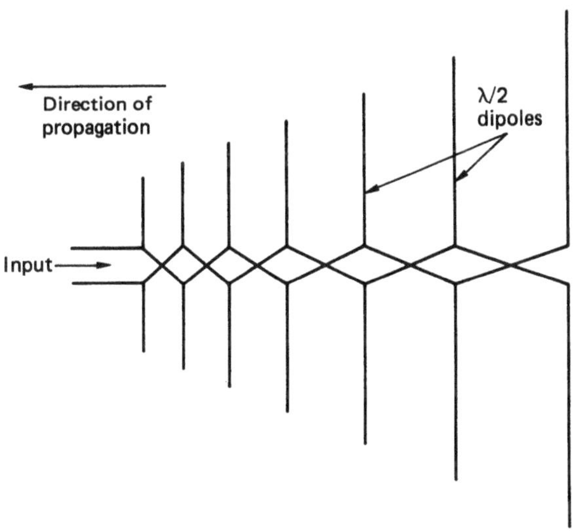

Figure 3.7 Basic half-wave transposed dipole LPA

excited and form an 'active region' that radiates energy. As the frequency is increased, the active region moves along the antenna from the longest to the shortest dipole and the inter-dipole spacings are chosen so that the radiation pattern remains the same and independent of frequency. The range of frequencies that can be covered by an LPA is very large and is only limited by the size that is acceptable for the longest dipoles. In practice, LPAs covering the whole HF broadcast band from 6 MHz to 26 MHz are quite common.

Because the active region moves from the longest to the shortest dipoles, with increasing frequency, it is possible to mount an LPA so that the electrical height remains constant or, alternatively, varies in a prescribed way with frequency. This is illustrated in Figure 3.8(a) for the case of constant electrical height and in Figure 3.8(b) for the case of uniform variation of electrical height. If the electrical height of an LPA remains constant, the TOA remains constant, and the antenna is therefore suitable for operation over a fixed path length throughout the diurnal, seasonal and sunspot cycles. On the other hand, a uniform variation of electrical height will give a variation of TOA so that a number of path lengths can be covered.

In the basic LPA shown in Figure 3.7, the dipoles are shown as conventional rods and such an arrangement is often used in practice. However, because of the long wavelength, rods are mechanically inconvenient in the HF band and wire dipoles with a sawtooth shape offer a greater bandwidth which increases the gain and power-handling capacity and reduces the VSWR.

The gain of a single LPA, mounted over an earth plane, is in the range 10–12 dBi and, to achieve higher gains, it is necessary to operate an array comprising a number of LPAs. Arrays are available comprising from two up to four stacked LPAs, the latter giving a gain of 22 dBi (comparable to the gain of an HRS 4/4/h curtain antenna). At the power levels required for HF broadcasting the electrical fields on an LPA are very high and, to minimize the radiator-tip voltages, cylindrical cages of sawtooth arrangements of wires are used. Despite

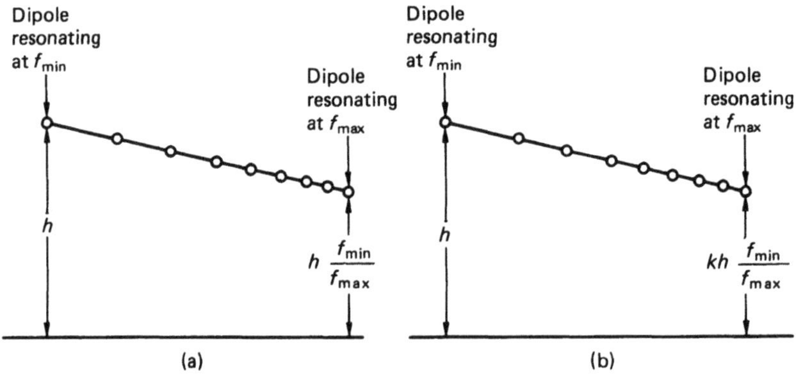

Figure 3.8 Control of electrical height of LPA. (a) Constant electrical height. (b) Uniform change of electrical height by factor *k*

these precautions, however, the power rating of an LPA array is limited although it increases with the number of basic LPAs used. Well-designed LPA arrays can handle up to 500 kW carrier power with 100% modulation. The VSWR of an LPA array is generally less than 1.5 : 1 over the design bandwidth and is suitable for operation with current designs of transmitting equipment.

The LPA array is very suitable for a 'dedicated' broadcast operation where the objective is to serve a single target area. This is because the wide bandwidth enables the whole range of transmission frequencies required for the diurnal, seasonal and sunspot cycles to be carried by a single antenna. However, for general broadcast use, where it is required to serve target areas at different distances, and distributed over a range of bearings, the LPA array is less suitable than the curtain array because:

1. Particularly for arrays comprising only a few LPAs, the power-carrying capacity is limited.
2. It is difficult to provide an azimuth-slew facility so that the range of bearings that can be served is limited.
3. It is not possible to provide for vertical slew, which is necessary if coverage is to be provided at a range of distances.

3.3.5 Miscellaneous wideband HF antennas

Over the years many different types of wideband HF broadcast antennas have been used but they can all be considered as falling into one or other of the three categories of broadside arrays, end-fire arrays and travelling-wave antennas. In broadside arrays the individual radiating elements are co-phased and arranged in a line, or plane, at right-angles to the direction of the service area. In end-fire arrays the radiating elements are arrayed in a line and driven with currents whose phases give maximum signal in the direction of the line of elements. Travelling-wave antennas take advantage of the fact that, if a wave is propagated along a wire, a radiation lobe is generated in approximately the direction of the wave.

Many different types of broadside HF broadcast arrays have been used over the years, including, in particular, the 'Marconi', 'BBC' and 'Sterba'. However, these arrays had limited bandwidth, slew capability and power rating, and have been superseded by the type of wideband, wide-slew array of horizontal dipoles described in Section 3.3.3. Similarly, a number of end-fire arrays have been used for HF transmission but, until the development of the LPA, the problem of achieving the optimum current phases in the radiating elements proved difficult to solve. As a result, the LPA is virtually the only form of end-fire array now used for HF broadcast transmission.

The most commonly used form of travelling-wave antenna is the rhombic, which is shown in Figure 3.9 and is, in effect, a wide-spaced transmission line with a matched termination so that a unidirectional wave is set up by the

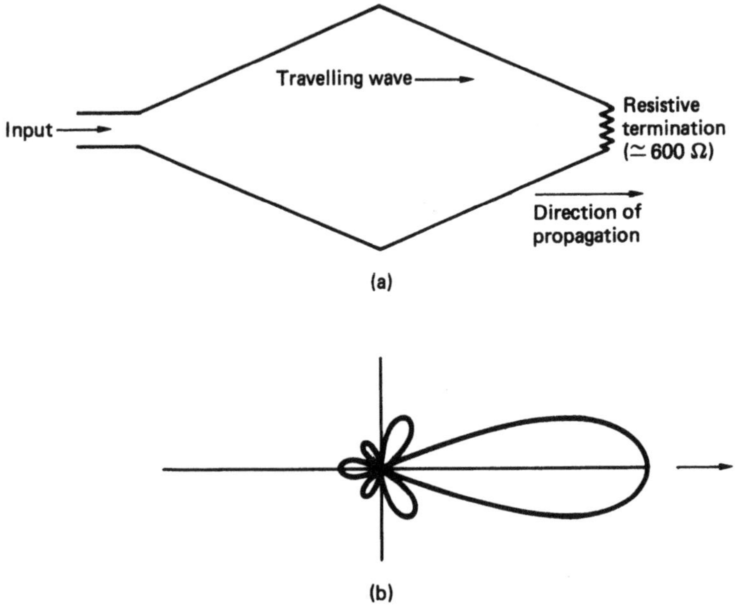

Figure 3.9 The rhombic antenna. (a) Antenna configuration. (b) Radiation pattern

transmitter. The rhombic is simple, cheap and easy to erect and, for these reasons, is still used for point-to-point HF communication links. But for broadcast transmission it suffers from a number of serious disadvantages:

1. The gain is relatively low, partly due to the loss of power in the resistive termination.
2. Unlike the curtain and LPA, the radiation pattern varies with frequency in an uncontrolled manner and it is impossible to achieve a constant performance over a wide frequency.
3. It is impossible to slew the main beam either in azimuth or in elevation.
4. The ground area required is much greater than that for a curtain or LPA.

Summarizing, as a result of the disadvantages of other types of antenna, the great majority of HF broadcast transmitting antennas in current use are either dipole curtains or LPAs.

3.3.6 Feeder switching and matrix arrangements

A large HF transmitting station will be required to serve target service areas distributed over a wide range of azimuth angles, and at widely differing distances. Further, it will also be required to provide services to more than one target service area simultaneously. The station will therefore be equipped with a number of transmitters (equal to the number of simultaneous transmissions plus, possibly, a spare transmitter) and a number of antennas (to cover all the azimuth angles, distances and frequencies required to serve the target areas over the diurnal, seasonal and sunspot cycles). In general, the number of antennas will be greater than the number of transmitters, and to give complete operational flexibility, it is desirable to connect any transmitter to any antenna. This comprehensive connection of transmitters is achieved by an RF switchbay (usually referred to as 'the matrix') which is illustrated, schematically, in Figure 3.10. It will be seen that the matrix is interposed in the transmission chain between any transmitter and any antenna and, as a result, it must meet stringent performance requirements:

1. The matrix design must be 'modular' so that additional switch units can be added at any time if the number of transmitters and/or the number of antennas is increased.
2. The power rating of each switch must be adequate to handle the peak and average powers of a transmission which may be as high as 2000 kW peak and 750 kW average.
3. There will be coupling between different transmissions carried simultaneously by the matrix. To avoid excessive cross-talk between transmissions the coupling between any two channels must not exceed −55 dB, measured on the matrix.
4. Unavoidable discontinuities in the matrix switches will increase the VSWR at the output terminals of each transmitter. This increase in VSWR will, of

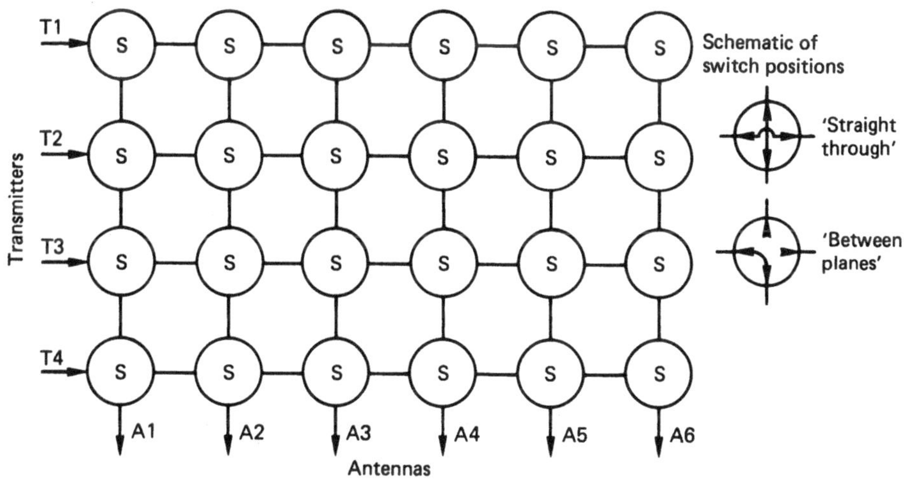

Figure 3.10 4 × 6 transmitter antenna switching matrix. The transmitter feeders are arranged to be in one plane (rows) and the antenna feeders in another parallel plane (columns). At each 'cross-over' there are two coupled switches – one in each plane. In practice the planes are vertical for coaxial matrices and usually overhead horizontal for balanced screened feeder matrices

course, increase with the number of switches interposed, but will tend to a maximum dependent on the number of switches and their relative physical separation. To limit the increase to an acceptable value the matrix must not introduce a VSWR exceeding 1.3:1 when terminated in a matched load.
5. The changeover time for the re-routing of a transmitter to a different antenna must be less than the frequency-changeover time of the transmitter (typically less than five seconds).

It is important to remember that virtually all HF transmitters have an unbalanced output (normally 50 Ω) whereas virtually all HF antennas have a balanced input (usually 300 Ω. It is obvious, therefore, that a 'balun' (an unbalanced to balanced transformer) must be included in the transmission chain between the transmitter and the antenna. As a result, a number of feeder switching arrangements are possible:

1. The balun can be placed at the transmitter output and screened balanced feeders and a matrix comprising screened balanced switches used to connect the balun output to the antennas.
2. A matrix comprising unbalanced coaxial switches can be used and baluns placed at the antenna outputs. In this case, coaxial feeders are used between the transmitters and the matrix and balanced screened or open-wire feeders between the matrix and the antennas.
3. A matrix comprising unbalanced coaxial switches can be used and baluns placed at the input terminals of the antennas. In this case, coaxial feeders are used between the transmitters and the matrix and between the matrix and the antennas.

It will be seen, therefore, that the matrix can be of balanced design (based on the use of 300-Ω screened balanced switches) or, alternatively, of coaxial design (based on the use of 50- or 75-Ω coaxial switches). Both these types of matrix can meet the performance specification outlined above but they differ in a number of important respects:

1. Balanced switches and open-wire transmission lines are significantly cheaper than coaxial switches and coaxial transmission lines.
2. Balanced switches are large and the matrix cannot be installed in the transmitter building. On the other hand, coaxial switches are compact and can be conveniently accommodated in the transmitter building with consequent advantages as regards security.
3. Open-wire lines and balanced switches are less reliable than coaxial types but they can be repaired rapidly with simple tools. On the other hand, the repair of coaxial lines and switches requires special tools and specially trained staff and can take some hours to complete.

It will be seen that either a balanced line or a coaxial matrix is acceptable and the choice between the two will depend on local circumstances. If security is of prime importance then the advantage will lie with the coaxial type, but if it is necessary to keep costs down to a minimum a balanced system will be attractive.

Chapter 4

VHF and UHF propagation and antennas

4.1 Introduction

Until the 1930s radio broadcasting used frequencies ranging from the LF band (153–281 kHz) through the MF band (520–1650 kHz) to the HF band (2–26 MHz). However, in the late 1930s the advent of television (which requires a high modulation bandwidth) made it necessary to bring the higher transmission frequencies of the VHF Band I into use. It is no exaggeration to say that technical problems increase as the transmission frequency is increased but, over the period from the late 1930s to the early 1960s, technical advances enabled frequencies in the VHF Bands II and III and the UHF Bands IV and V to be brought into use. However, it should be noted that, due to decreasing range and 'patchy' coverage, frequencies above the UHF bands (i.e. above 1000 MHz) are not suitable for wide-area broadcast coverage and are only used for local services where line-of-sight transmission is possible.

In addition to making high-definition television transmissions possible, the use of frequencies above the HF band has enabled the high-quality VHF/FM radio transmission systems to be introduced. In the VHF/FM system, the programme material modulates the carrier frequency (unlike the MF/AM system where the carrier amplitude is modulated) and this makes possible:

1. A high signal-to-noise ratio (SNR) which results from the use of frequency modulation. In effect, the FM system exchanges an increased channel width for improved SNR.
2. A higher audio frequency response (15 kHz as opposed to the 6 kHz possible with channel allocations in the MF band), which gives 'high-fidelity' reception.
3. More efficient use of channel allocations resulting from reduced susceptibility to co-channel interference.
4. Stereophonic transmission that requires the transmission of two audio signals (left and right hand) and hence requires a higher modulation bandwidth than can be used in the LF and MF bands.

The bands now in use for television and VHF/FM broadcasting vary somewhat from country to country but, in general, are as summarized in Table 4.1.

Table 4.1

Band	*System*	*Frequency range (MHz)*	*Transmission*
I	VHF/TV	47–69	Colour televison
II	VHF/FM	88–108	Stereo radio
III	VHF/TV	170–250	Colour television
IV	UHF/TV	470–600	Colour television
V	UHF/TV	600–860	Colour television

4.2 Propagation

The propagation of VHF and UHF transmissions is quite different from that of MF or HF transmissions. Ground conductivity and ionospheric propagation have little or no effect and the principal factors governing attenuation of the transmitted signal are earth curvature, shadowing by natural and man-made obstructions and ground reflection.

4.2.1 Propagation fundamentals

To a first approximation, in the VHF and UHF bands the ground can be considered as a perfect reflector and it is convenient to consider, first, transmission over a flat earth. This case, which is illustrated in Figure 4.1, is applicable to transmission over a distance short enough for the earth's curvature to be ignored (less than about 10 km). As shown in Figure 4.1, the signal reaches the receiving antenna by two paths: directly and after reflection from the earth plane which reverses the phase. This phase reversal means that, when the difference in the lengths of the two paths is small, their contributions tend to cancel and the amplitude of the received signal is low. The received field strength (except at distances of less than 1 km from the transmitter) can be shown to be:

$$e = 9.34 f P^{1/2} h_T h_R / d^2$$

where e is the field strength in uV/m (field strength is, however, normally quoted in dBμ, i.e. 20 log (e); see Appendix 2), f is the frequency in MHz, P is the ERP (effective radiated power) in kW (ERP with antenna gain relative to a half-wave dipole; see Appendix 2), h_T is the transmitting antenna height in m, h_R is the receiving antenna height in m, and d is the path distance in km. It will be seen that, unlike MF or HF transmission, the field strength is directly proportional to the heights of the transmitting and receiving antennas, which therefore have an important effect on the extent of the service area. For example, if the antenna heights are both doubled the field strength is multiplied by 4 (an increase of 12 dBμ).

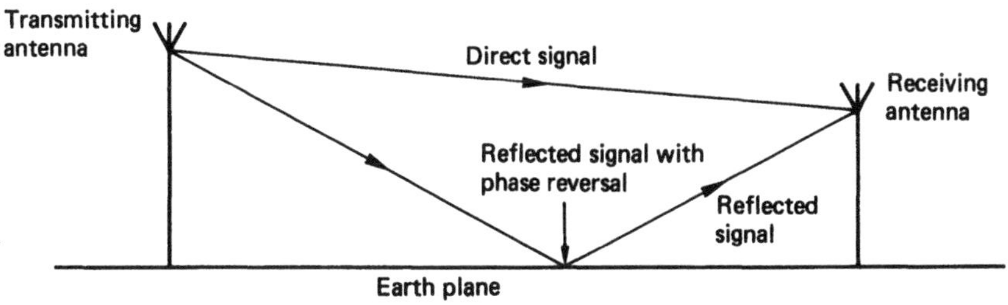

Figure 4.1 Transmission over flat earth

In practice, the range of coverage of a VHF or UHF transmitting station will be such that the curvature of the earth has an effect on propagation, and the above expression must be multiplied by a factor that takes account of this. This factor depends on the path length and antenna heights, as the latter control the location of the 'radio horizon' (the point at which the path between the transmitting and receiving antennas ceases to be line-of-sight). Figure 4.2 illustrates the location of the radio horizon and the obstruction of the line-of-sight path due to earth curvature on long paths. At receiving sites near the transmitter the field strength can exceed the 'free-space' value (owing to reinforcement of the direct and earth-reflected signals) but at distances beyond the radio horizon it decreases rapidly. The higher the frequency the greater is the attenuation beyond the radio horizon and this is due to the increasingly 'optical' nature of the propagation. It follows that, particularly in the UHF band, the use of a higher transmitter power may not give a worthwhile increase in coverage.

So far it has been assumed that the surface of the earth is smooth (i.e. ground irregularities and obstructions are small compared with a wavelength). In practice this is not true and, due to the 'shadowing effect' of ground irregularities and natural and man-made obstacles, field strengths are lower than those obtaining over a smooth earth. This is particularly true in the UHF band where it is necessary to use ERPs up to 10 dB higher than those that give a satisfactory service in the VHF bands.

Although VHF and UHF transmissions are not reflected from the ionosphere, fading and co-channel interference can occur due to 'tropospheric propagation'. The troposphere is the region 1–10 km above the earth (where clouds form) and

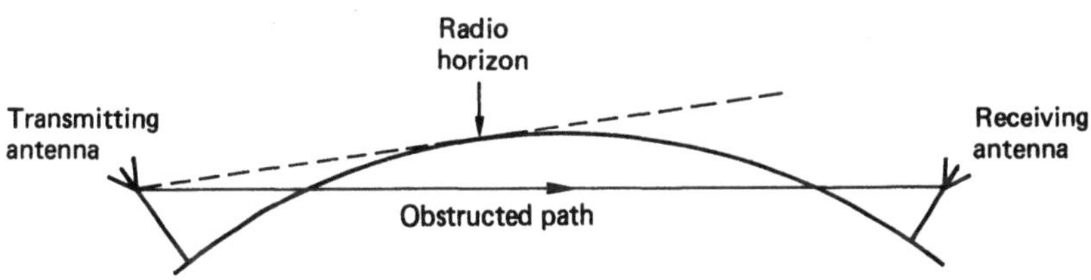

Figure 4.2 Effect of earth curvature

where, due to a decrease of atmospheric permittivity with height, VHF and UHF signals are refracted and returned to earth. Tropospheric signals can be significant at distances of about 80 km and above from the transmitter and result in fading which, particularly in the case of television transmissions, limits the potential service area. Under stable weather conditions, 'ducts' can form between layers of air in the troposphere so that relatively strong signals can be propagated for distances of hundreds of kilometres, causing co-channel interference. Such interference is usually seen as a pattern of sloping light and dark bands, but occasionally as a ghostly superimposed image.

4.2.2 Polarization

In the expression for field strength in Section 4.2.1 it is seen that the field strength decreases with decreasing receiving antenna height, becoming zero at ground level. In practice, this is true for horizontal polarization (HP) but not for vertical polarization (VP). In fact, the VP field strength is similar to the HP field strength except at heights around 10 m or less above ground level. At these low heights, the VP field strength does not decrease to zero at ground level (as does the HP field strength) but remains approximately constant at the value occurring at about 10 m above ground level.

For television broadcasting, the difference between VP and HP field strength at low heights above ground level is relatively unimportant as, to achieve good ghost-free reception, it is common practice to use receiving antennas mounted at a height in the region of 10 m (e.g. on the roof of a house), but, in any event, as viewing requires the full attention of the individual, there is little demand or practical application for portable or mobile reception. However, the position as regards VHF/FM reception is very different, as portable radios and car radios (with simple built-in 'whip' antennas) are in common use. Car radios (and most portable radios) are operated at low heights above ground level and VP transmission is therefore to be preferred to HP transmission, since it gives a greater field strength and, more importantly, can be received on vertical 'whip' antennas which have omnidirectional reception characteristics in the horizontal plane. Originally in most countries VHF/FM transmissions used HP because, prior to the invention of the transistor, the proliferation of portable and car radios was not anticipated, whereas the directional characteristics of simple fixed 'roof-top' HP receiving antennas were expected to minimize the risk of impulsive electrical and co-channel interferences. Experience has shown that interference has been less of a problem than expected, partly as a result of legislation requiring the suppression of spurious radiation from petrol engine ignition systems and other electrical equipment; however, when the re-engineering of existing services is undertaken it is now common practice to transmit a signal which comprises both VP and HP components, which may be 90° out of phase, giving CP (circular polarization), or in phase, giving SP (slant polarization). In this way existing HP receiving installations can still be used whilst the advantages of VP for portable and mobile reception is introduced. However, when VHF/FM services are being established in a previously unserved area the optimum choice is almost certainly VP. Costs will be minimized: compared with CP, VP antennas are less complex, impose lower wind loading on the antenna support structures, and transmitters of only half the power, having reduced operating costs, are required. CP, when used with fixed HP receiving installations, may offer a marginally better service than VP but there is little or no evidence of this, nor is there evidence to suggest that in practice CP is better than SP, although the former would appear to have some theoretical advantages. CP or SP have also sometimes been used for television transmissions when it has been thought necessary to cater for viewers using simple loop or 'rabbit ear' antennas at a low height above ground, but the use of such polarizations results in increased co-channel interference or channel allocation constraints.

As mentioned previously in the case of television, co-channel interference, due to transmissions from a distant transmitter propagated via the troposphere, can result in obtrusive 'herringbone' patterns on the viewer's screen. This interference can be reduced substantially (by up to 10 dB) if different polarizations are used at adjacent co-channel stations (i.e. VP at one station and HP at the other). A further reduction of about 10 dB in the visibility of interference can be obtained by making the vision carrier frequencies of co-channel stations differ by an accurate multiple of the line-scan frequency. This is known as 'offset operation' and offsets of one-half and two-thirds of the line-scan frequency accurate to ±500 Hz are commonly used. If the frequencies are controlled to ±1 Hz, this is known as 'precision offset' working and the interference is reduced by a further 10 dB, but the equipment to do this is expensive.

4.2.3 Required field strengths

The field strength required to provide a VHF/FM service or a VHF or UHF television service depends on the noise factor of the receiver, the gain of the receiving antenna and the incidence of man-made interference. Taking these factors into account, the minimum field strengths shown in Tables 4.2 and 4.3, which are based on average conditions as regards receiver performance, antenna

Table 4.2 (VHF/FM minimum required field strength in dBμ)

Band	*System*	*Rural areas*	*Suburban areas*	*City areas*
II	Mono	48	60	70
II	Stereo	54	66	74

gain and noise, can be used for planning purposes. These field strengths are the averages that must be achieved in any area in order to provide a service. If they are achieved it does not mean that every listener or viewer will receive a high-grade service. It does mean, however, that the majority (90% or more) should receive an acceptable service.

These field strengths assume that a receiving antenna mounted at a height of 10 m above ground level is used. If a receiver with a built-in whip antenna is used at ground level, the stereo field strengths above will give good mono reception. The stereo coverage will, however, be very restricted.

Table 4.3 (Television minimum required field strength in dBμ)

Band	*System*	*Grade I service*	*Grade II service*
I	PAL/NTSC	70	55
III	PAL/NTSC	75	60
IV	PAL/NTSC	80	65
V	PAL/NTSC	85	70

A Grade I service is one in which interference is imperceptible and a Grade II service is one in which, although interference is perceptible, it is not disturbing to the viewer.

4.2.4 Propagation analysis

As in other bands, the object of propagation analysis is to forecast the received field strength of a proposed installation in terms of transmitter power, antenna height, antenna gain and other relevant parameters. Alternatively, the object may be to arrive at a set of transmitting-equipment parameters that will achieve a required service area.

In all frequency bands, the received field strength depends on the path length and the ERP (Appendix 2). However, as in other bands, field strength also depends on what may be called the 'path parameters' and these vary with the frequency band. In the LF and MF bands the path parameters are the path length, ground conductivity, ground permittivity and the curvature of the earth. In the HF band, where sky-wave transmission is normally used, the path parameters are the path length and the characteristics of the ionosphere. In the VHF and UHF bands, path parameters are the path length, the transmitting and receiving antenna heights, the nature of the terrain between transmitting and receiving antennas and the curvature of the earth.

(a) Field-strength prediction curves

In the VHF and UHF bands, received field strength depends on:

1. Path length
2. ERP (effective transmitter power)
3. Transmitting antenna effective height
4. Receiving antenna effective height
5. Loss due to obstructions and earth curvature

Factors (1) and (2) present no difficulty as they are known, or are assumed for the purposes of calculation. The difficulty of VHF/UHF propagation analysis is concerned with the calculation of factors (3) to (5), which depend on the nature of the path profile (i.e. a cross-section of the terrain between the transmitter and receiver). This is because propagation in the VHF and UHF bands is somewhat similar to that of light rays, in that obstructions and earth curvature create shadow areas in which the received signal arrives by diffraction and, consequently, is reduced. Further, the effective heights of the transmitting and receiving antennas are measured above the average height of the surrounding terrain and will, in general, differ from their heights above ground, measured at the transmitting and receiving sites. The field strength at any point in the service area, therefore, can only be determined accurately by relatively complex calculations based on an analysis of a detailed path profile. These calculations present no difficulty for point-to-point radio and television links where only a single path length is of interest. But in the case of a broadcasting station, the labour involved in calculating the field strength at a large number of receiving points may be prohibitive or, alternatively, require a large computer-based store of terrain information.

Because of the difficulty involved in calculating the complete coverage of a broadcasting station, the FCC and CCIR have published sets of curves giving VHF and UHF field strengths over average terrain in terms of path length and effective transmitting antenna height. The curves assume a receiving antenna height of 10 m and an ERP of 1 kW and have been derived by averaging measurements of a large number of broadcast transmissions. The curves give the average field strength exceeded for 50% of the time at 50% of locations and, for this reason, are sometimes referred to as '50/50 curves'. Three sets of curves are available and are for 45–108 MHz (the VHF Bands I and II), 175–212 MHz (the VHF Band III) and 460–850 MHz (the UHF Bands IV and V). It must be stressed that the accuracy of the field strengths provided by the curves cannot be high. For rough, obstructed paths the values tend to be optimistic (high) and for smooth, unobstructed paths they tend to be pessimistic (low). Despite this, however, the curves are useful for the preliminary assessment necessary when a radio or television transmitting station project is planned.

To facilitate calculation, the empirical formulae on which the curves are based have been incorporated in computer programs for the three frequency

VHF BANDS, I, II & III FIELD-STRENGTH PREDICTION

FREQUENCY RANGE:	BAND III
TRANSMIT ANTENNA EFFECTIVE HEIGHT:	100.000 m
RECEIVE ANTENNA HEIGHT:	10 m
EFFECTIVE RADIATED POWER:	80.000 kW

DISTANCE, KM	F/S DBU
10.000	92.000
20.000	80.000
30.000	72.000
40.000	66.000
50.000	61.000
60.000	56.000
70.000	52.000
80.000	47.000
90.000	43.000
100.000	39.000

UHF BANDS IV AND V FIELD-STRENGTH PREDICTION

TRANSMIT ANTENNA EFFECTIVE HEIGHT:	100 m
RECEIVE ANTENNA HEIGHT:	10 m
EFFECTIVE RADIATED POWER:	80 kW

DISTANCE, KM	F/S, DBU
10.00	90.0
20.0	75.0
30.0	65.0
40.0	58.7
50.0	52.3
60.0	48.5
70.0	44.6
80.0	40.7
90.0	36.9
100.0	33.0

Figure 4.3 VHF and UHF field-strength prediction programs

bands (i.e. the VHF Bands I and II, the VHF Band III and the UHF Bands IV and V). Examples of the printouts from these computer programs are given in Figure 4.3.

(b) Calculation of field strength

As has been pointed out, the calculation of field strength (as opposed to the rough estimate given by the prediction curves) requires a detailed path profile of the terrain between the transmitting and receiving sites. The first step in the calculation is, therefore, to plot a path profile from information derived from large-scale maps (a scale of 1:5000 if possible) in a format that expands the height scale so that the terrain can be accurately represented. The format assumes that the radius of the earth is greater than its actual value (usually a value of 4/3 is used) to take account of the curved path followed by a radio beam

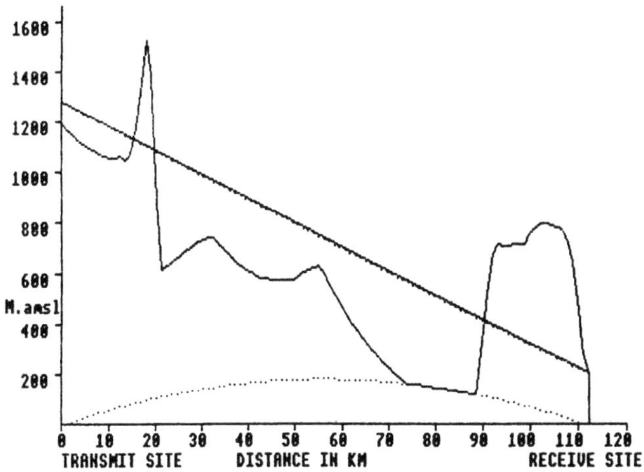

Path profile for path KEF to ANNABA

Height of transmitting antenna:	85 M AGL
Height of receiving antenna:	10 M AGL
Effective earth radius:	1.333

Figure 4.4 Example of earth profile

due to the change of the dielectric constant of the atmosphere with height. An example of a path profile is shown in Figure 4.4 and the use of a 4/3 earth radius enables the direct signal path between the transmitting and receiving antennas to be plotted as a straight line.

There are a number of methods in use for the calculation of field strength over a given path and that due to Bullington illustrates the approach. In this method, the calculation is carried out in three steps:

1. The received field strength is first calculated on the assumption that there is an unobstructed free-space path between the transmitting and receiving antennas. This field strength is given by:

$$E_{FS} = 10 \log (\mathrm{ERP}/d^2) + 107 \text{ dB}\mu$$

where ERP is the effective radiated power in kW, and d is the path length in km.

2. The loss due to the presence of a flat, smooth, unobstructed earth between the transmitting and receiving antennas is determined. This loss is given by:

$$L_{FE} = 20 \log [2 \sin (A)] \text{ dB}$$

where $A = (0.0024 h_T h_R f)/d$ degrees, h_T is the effective height of the transmitting antenna in m, h_R is the effective height of the receiving antenna in m, f is the frequency in MHz, and d is the path length in km. This loss takes account of the effect of signals reflected from the terrain between the transmitting and receiving antennas. In effect, the actual terrain is replaced by a flat, smooth earth at an 'average' height above mean sea level. However, the effective heights of the transmitting and receiving antennas above the smooth earth depend on the detailed nature of the terrain and, consequently, are difficult to determine. A number of empirical methods for determining the effective antenna heights are derived from an analysis of actual measurements over a wide range of terrain types. For example, the CCIR method assumes that the effective height of the transmitting antenna is its height over the average terrain height from 3 km to 15 km from the transmitter and the receiving antenna height is 10 m.

With the CCIR method, the formula for the loss L_{FE} assumes that the angle of incidence of the transmitting signal at the surface of the earth is low (i.e. that the earth behaves as a perfect conductor). The formula, therefore, is unreliable at distances within a few kilometres of the transmitting antenna and intended for the determination of the limit of coverage of a transmitting station. Calculation of the field strength in the region near to the transmitting station needs to take account of the ground conductivity, ground permittivity, polarization and vertical radiation pattern (VRP) of the transmitting antenna and is considered further in Section 4.3.3.

3. The loss due to obstructions in the line-of-sight path between the transmitting and receiving antennas (including the effect of earth curvature) is determined from the theory of diffraction over a knife-edge obstruction. The diffraction loss over a single knife-edge obstruction is shown in Figure 4.5 as a function of a variable V which depends on the height of the obstruction and its distance from the transmitting and receiving antennas. To determine the diffraction loss the actual obstructions are replaced by a single knife-edge obstruction, and the procedure for this is illustrated in Figure 4.6(a). Tangents are drawn from the transmitting and receiving mast

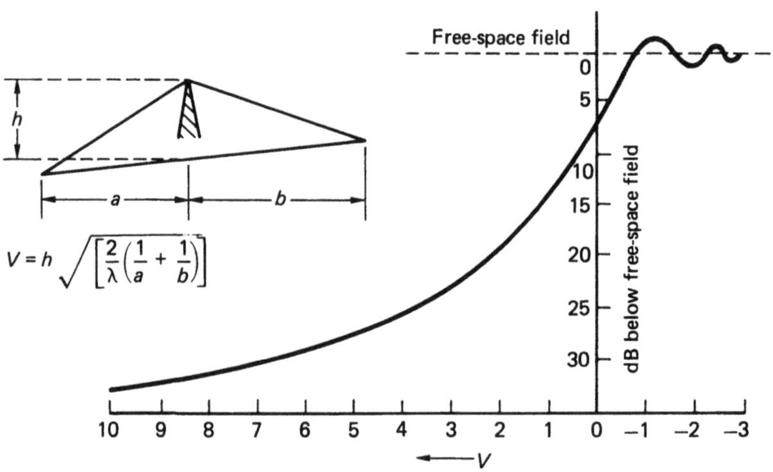

Figure 4.5 Loss due to diffraction over a knife edge

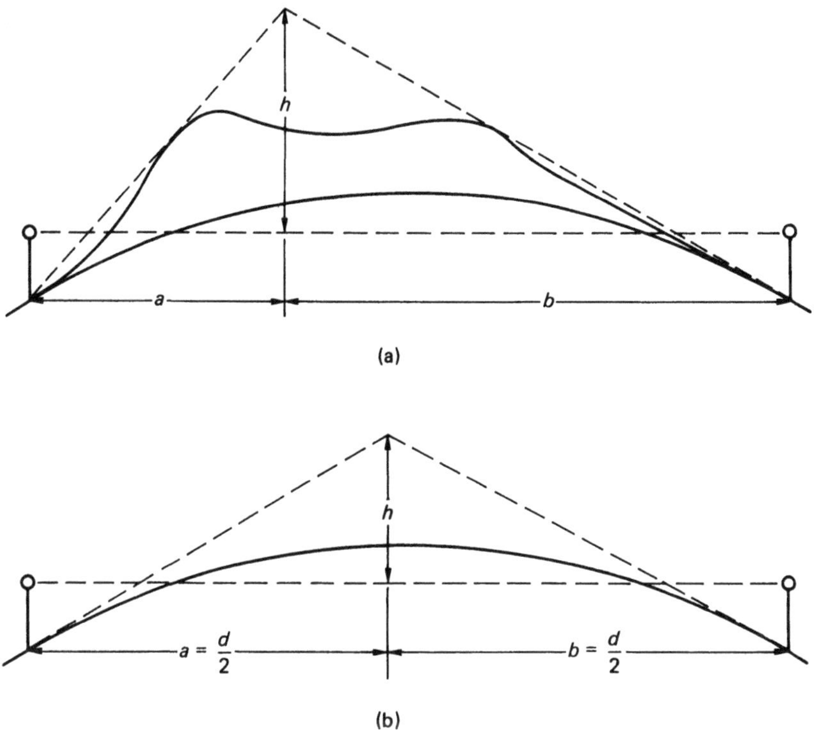

Figure 4.6 Determination of equivalent knife edge obstruction. (a) Obstructed path over curved earth. (b) Smooth path over curved earth. h = height of equivalent knife edge over line-of-sight-path

bases to (but not through) the obstructions and will intersect at a point higher than the top of the highest obstruction. The obstruction loss in dB can then be calculated from Figure 4.5 and is denoted by L_{OB}. If there are no actual obstructions there can be an effective obstruction formed by the curvature of the earth as illustrated in Figure 4.6(b) and, once again, the loss can be obtained from Figure 4.5.

In the case of point-to-point communication, the receiving antenna will be sited so as to be clear of local obstructions and the field strength can be calculated by following steps (1), (2) and (3) above. However, in the case of broadcasting, there will, in general, be an additional loss due to buildings and other local obstructions. The loss, in dB, which is denoted by L_{LO} depends on frequency and the receiving environment and can vary between about 3 dB in an open rural area through 6–10 dB in a suburban area to 20 dB or more in a city area.

The calculated field strength is determined from steps (1) to (3) above and is given by:

$$F = E_{FS} - L_{OB} - L_{LO}$$

Computer programs are available that facilitate the rapid calculation of steps (1) to (3) above. However, the calculation of field strength is still a laborious process since a ground profile must be prepared for each reception point considered. In most of these programs, the path profile is entered in the form of a number of terrain heights above mean sea level. The program then generates a large number of equal-interval terrain heights (typically 100) by interpolation so that the best possible approximation to the actual profile is achieved. The accuracy of the interpolated profile obviously depends on the number of heights entered which, in turn, depends on the scale of the maps available.

4.2.5 Interpretation of propagation analysis

As has been pointed out, field-strength prediction curves can only give a general guide to the coverage of a proposed transmitting station. Further, methods of calculation based on terrain profiles give the median field strength in a particular region in the service area of the transmitter. This means that a measurement at a particular point in the service area will not give precisely the calculated field strength at that point. It means, in fact, that in an area centred on the point considered, approximately 50% of locations (i.e. 50% of listeners or viewers) will receive at least the calculated field strength and approximately 50% will receive less than the calculated field strength. In many cases, however, it is necessary to estimate the median field strength required to ensure that most locations (say 90%) receive a particular value of field strength. The variation of field strength will depend on the receiving environment (rural, suburban or city) and can be expressed in median field strength required to ensure a service to 90% of locations (see Table 4.4).

Table 4.4

Standard deviation (dB)	*Increase in field strength (dBμ)*
0	0
3	4
6	8
9	12
12	16

For example, if the calculated median field strength in an area is 66 dBμ, then only 50% of listeners/viewers can expect to receive at least this value. Suppose now that the standard deviation of field strength variation is 6 dB (a typical value). In this case it will be necessary to increase the median field strength of $66+8=74$ dBμ to ensure that 90% of listeners/viewers receive a field strength of at least 66 dBμ.

4.3 Antennas

4.3.1 Antenna design considerations

As has been seen, the field strength of VHF and UHF transmissions and the area which can be served increases with increasing antenna height, and transmitting antennas are therefore, in general, mounted at the maximum practicable height above the terrain to be served, typically on a structure on top of a hill or mountain. The field strength at any point is also proportional to the square root of the power radiated by the antenna towards that point. Hence the radiation characteristics of the ideal antenna would be such as to provide adequate field strength over the maximum area, and at VHF frequencies, and to an even greater extent at UHF frequencies, it is practicable to go a long way towards meeting the ideal. The factor by which the field strength at the limits of the service area can be increased in this way, relative to a half-wave dipole, is defined as the gain of the antenna. Thus the gain of the antenna multiplied by the transmitter power is the ERP and it should be noted that, in general, increasing gain involves a lower initial overall capital cost and a very much lower revenue cost than an equivalent increase in transmitter power. For example, a system with an ERP of 100 kW could be achieved using a transmitter power of 10 kW and an antenna gain of 10 dBD (a power ratio of × 10) or, alternatively, with transmitter power of 50 kW and an antenna gain of 3 dBD (a power ratio of × 2). The capital and revenue costs of the system using a 10-kW transmitter power would be significantly lower than those of the system using a transmitter power of 50 kW.

To obtain a high gain, a VHF or UHF antenna normally comprises 'vertically stacked' radiating elements, driven with co-phased currents and mounted near the top of a mast or tower, as illustrated in Figure 4.7 for an 'eight-tier' antenna. The gain increases as the aperture of the antenna is extended but, with the top of the antenna at a given height, the mean height is reduced with a consequent reduction in field strength. Also, as the vertical aperture is increased, the beamwidth in the vertical plane is reduced and eventually there is a risk that 'mast-sway' will cause unacceptably large variations in the received field strength or they will be too low relatively near the transmitting site. These considerations put limits on the vertical aperture of VHF and UHF antennas of about 40 m in Bands I, II and III, 22 m in Band IV and 14 m in Band V.

The power gain of a stacked vertical array of radiating elements is approximately equal to the aperture in wavelengths. If, therefore, the maximum vertical apertures given above are assumed, the maximum gain obtainable will be approximately as shown in Table 4.5, although in practice if transmitters in Bands IV and V are sited in urban areas it is prudent to utilize antennas with gains not exceeding 13 dB to guarantee first class reception under all operating conditions (see Section 4.3.4).

Sometimes it is necessary to use a pair of feeders to carry the required power to an antenna, but irrespective of the need for this, experience has repeatedly

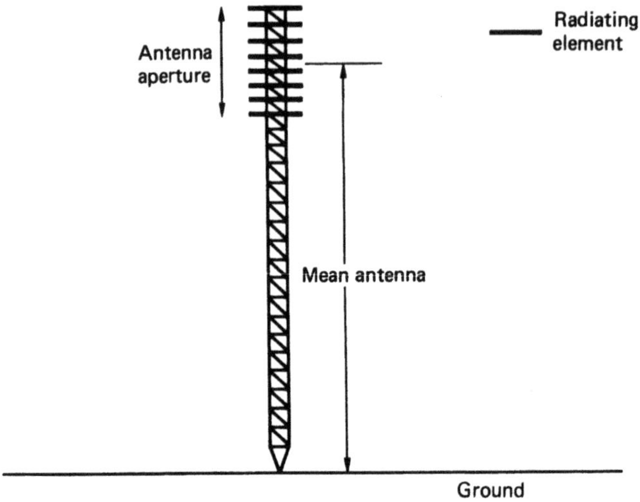

Figure 4.7 Eight-tier VHF/UHF antenna

shown that a VHF or UHF transmitting antenna comprising two halves independently fed is the preferred arrangement; particularly if the design permits through access for maintenance personnel to antennas or obstruction lights mounted higher up the support structure, or the maintenance of one half of the antenna whilst the other remains operative.

Typically a transmitting site will be established to provide one or more services with limited hours of broadcasting, but as time goes by the requirement for extended hours of broadcasting and new services in the same or other bands inevitably arises and eventually (after perhaps 25–35 years) the complete re-engineering of the antenna. It is at these times that 'split' antennas with generous safety factors and bandwidths which anticipated the future are invaluable.

Table 4.5

	Maximum gain	
Band	*Power gain*	*Gain (dBD)*
I	× 8	9
II	× 13	11
III	× 24	14
IV	× 40	16
V	× 40	16

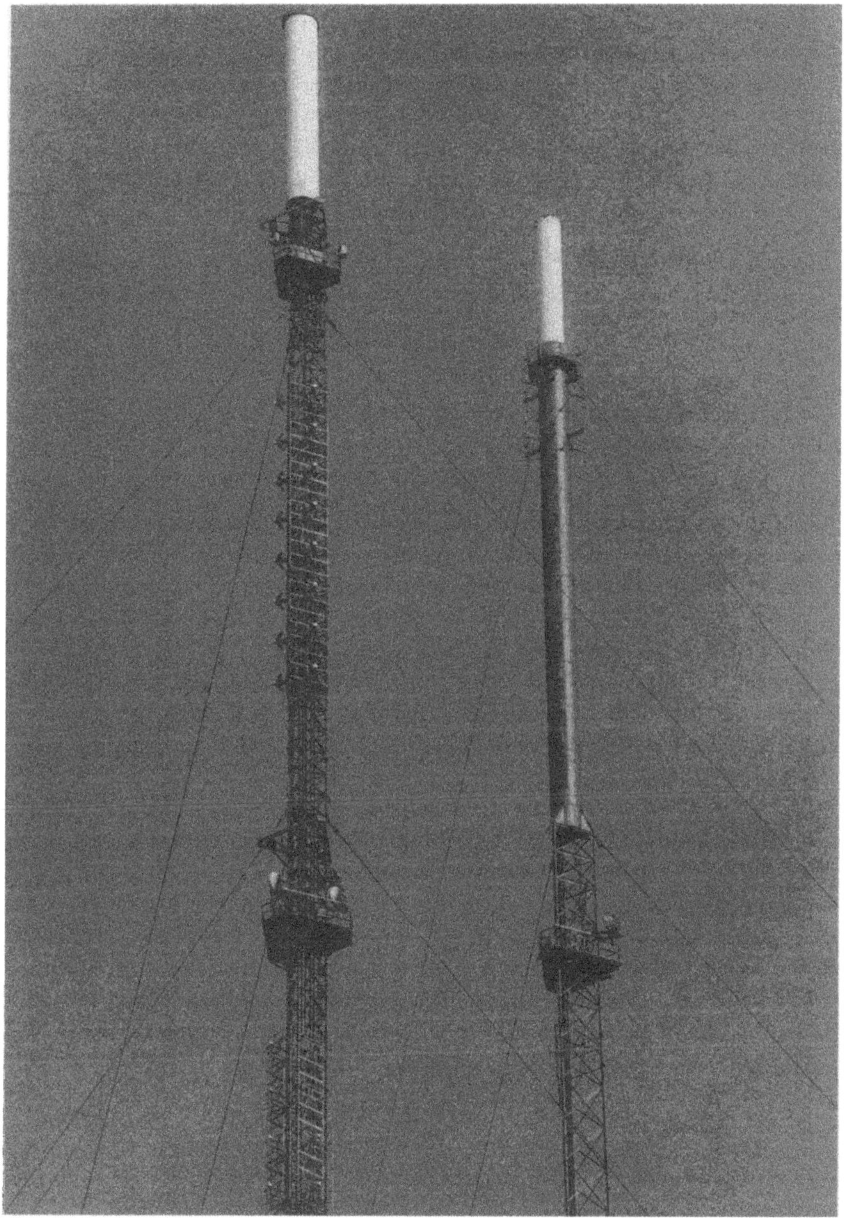

Figure 4.8 Re-engineering of VHF/UHF antennas. The mast on the right carrying a four-channel 500-kW ERP/programme UHF television antenna, a redundant two-tier band I television antenna and a four-programme 120-kW ERP horizontally polarized FM radio antenna is being replaced by a new structure carrying a similar UHF antenna, a five-programme 240-kW circularly polarized FM radio antenna and a directional FM radio antenna. When carrying out such work the relative positions of the masts must be chosen to minimize degradation of service due to shadowing and reflections from the adjacent structure (Courtesy of Engineering Information, BBC, London)

Whether or not the preferred arrangement is therefore specified will depend the funds available and the importance of the services, e.g. the extent to which breaks in service are acceptable or the size of the population affected. If a support structure is to carry a number of antennas operating in different bands it is generally desirable to put higher band antennas above those operating in lower bands, though when directional patterns are required or if different service areas are to be covered this may be less important.

4.3.2 Antenna types

(a) Support structure mounted antennas

In many cases a VHF/FM or VHF/UHF-TV antenna is assembled from a number of standard panels which each comprise a number of radiating elements mounted on a plane reflector. These panels are placed on the faces of a mast or tower (for convenience the term 'mast' will be used for both types) of triangular or square cross-section. The radiating elements of a panel are normally vertical or horizontal dipoles fed from,a single input via a branch feeder system (see Appendix 2). Using standard antenna panels, in different configurations, and carrying currents with varying phases, antenna systems with varying gains and radiation patterns can be provided. A basic panel is usually designed to have a directional radiation pattern with maximum radiation normal (i.e. at right-angles) to the panel. For a mast with square cross-section, the azimuth radiation pattern is designed to have a 90° 6-dB beamwidth as illustrated in Figure 4.10. By this means, in the diagonal directions, the fields from adjacent panels add to give the same field as in the directions of maximum radiation. For a triangular mast section, of course, the 6-dB beamwidth needs to be 120°. Panels designed on this basis, and fed with equal powers, allow omnidirectional azimuth radiation patterns to be achieved. The maximum/minimum ratio of the azimuth radiation pattern increases with the size of the mast, and Table 4.6 gives this ratio in dB as a function of the mast width in wavelengths.

Table 4.6

Mast face width	*Maximum/minimum ratio (dB)*	
(wavelengths)	*Square mast*	*Triangular mast*
<0.25	>0.1	>0.1
0.5	0.2	0.1
1.0	1.5	1.2
1.5	4.5	2.9

Figure 4.9 Rigid distribution feeders of a UHF antenna being assembled prior to testing (Courtesy of Alan Dick & Co, Cheltenham)

For example, at 600 MHz in the UHF band (a wavelength of 0.5 m) a square mast of 0.5 m width will result in a maximum/minimum ratio of 1.5 dB. In practice, a maximum/minimum ratio of up to about 4 dB is usually considered acceptable and this means that the face width of a square mast should not exceed 0.6 m at the top and 1.1 m at the bottom of the UHF band.

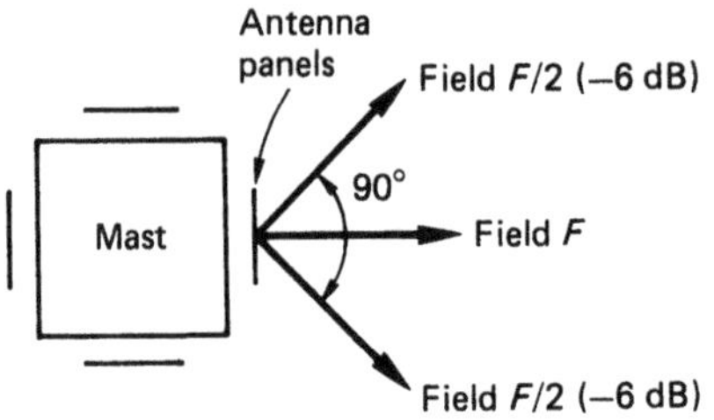

Figure 4.10 Panel with 90° 6-dB beamwidth

Panel antennas enable a directional radiation pattern to be readily achieved. A wide variety of azimuth patterns can be obtained by varying the magnitude and phase of the excitation of panels round the mast or using different numbers of panels on different mast faces.

The arrangement of vertically and horizontally polarized panels, including the feed system, is illustrated, schematically, in Figure 4.11. As has been explained in Section 4.2.2, VHF/FM transmissions with CP are required to increase the field strength at low heights above ground and cater for portable and car radio reception. To this end VHF/FM panels comprising both vertical and horizontal or crossed dipoles are available (Figure 4.12(a)).

The radiating elements of UHF panels, which are relatively small and fragile, are usually enclosed in a plastic cover, either individually or in a cylinder enclosing the complete antenna, to protect them and minimize the effect of icing.

Log-periodic elements may be used for low-power VHF and UHF antennas to provide horizontally, vertically or slant-polarized transmissions.

(b) Cantilever antennas

There is a wide variety of VHF and UHF transmitting antennas available that can be installed on narrow, unstayed, small-diameter poles, or structures, or

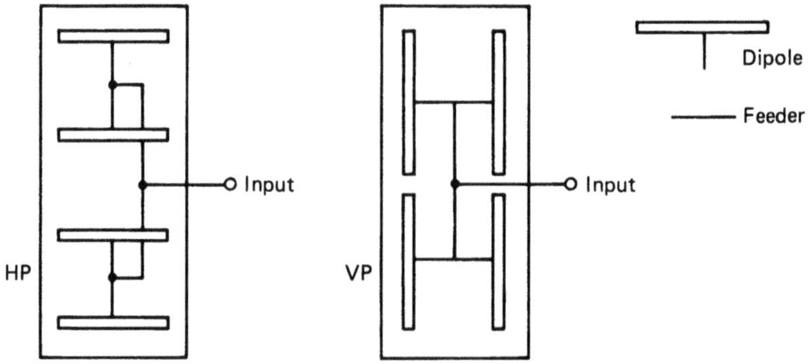

Figure 4.11 Horizontally and vertically polarized panels

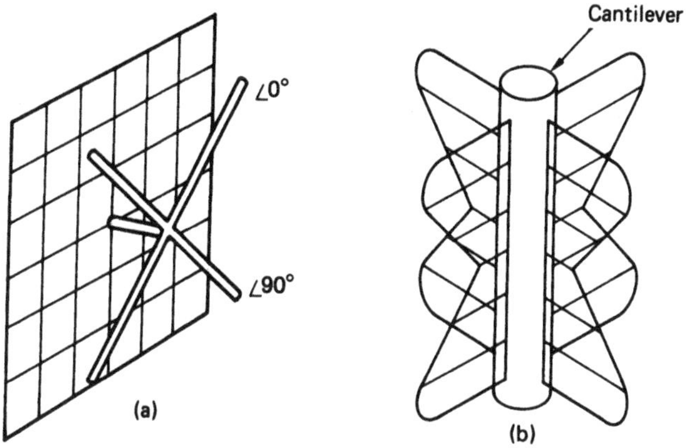

Figure 4.12 Cantilever-mounted antennas. (a) Crossed dipole panel (circular polarization). (b) Batwing (superturnstile) (horizontal polarization)

inside structural glass-reinforced plastic cylinders mounted at the top of masts (an arrangement often termed a 'cantilever' antenna). Some antennas of this type are supplied as a complete unit including the mounting pole and distribution feeder system but in other cases radiating elements are supplied that can be mounted on or in the cantilever. Figure 4.12(b) illustrates the batwing (superturnstile) which is supplied as a complete unit and is often used for UHF applications requiring an omnidirectional radiation pattern. Also shown is a 'crossed dipole' radiating element generating a CP transmission designed for Band II and Band III transmissions. A number of these elements can be mounted on a cantilever to form a high-gain antenna. Cantilever antennas normally involve a lower overall capital cost, taking into account support structure costs, than support structure mounted antennas, but do not offer the same maintenance flexibility.

4.3.3 Antenna performance parameters

The important performance parameters of a VHF or UHF antenna are the gain, VRP, azimuth (horizontal) radiation pattern, impedance, ERP and power-carrying capacity.

(a) Overall gain

Antenna gain has been discussed in Section 4.3.1 but it is important to note that, to obtain the 'effective gain', feeder loss must be deducted from the gain of the antenna itself. For example, a UHF antenna operating at 600 MHz (wavelength 0.5 m) and with a vertical aperture of 20 m will have a gain (in isolation) of

10 log (20/0.5) = 16 dBD. If the antenna is fed via 100 m of $\frac{7}{8}$-inch helical air-dielectric, semi-flexible feeder the loss will be approximately 3 dB and the effective gain of the antenna will be 16 − 3 = 13 dBD.

(b) Vertical radiation pattern (VRP)

If the stacks of a VHF or UHF antenna are driven with equal, co-phased, currents the VRP comprises a main lobe directed horizontally and a number of minor lobes directed above and below the horizontal. This is illustrated in Figure 4.13 where it will be seen that the nulls between the minor lobes will result in areas of low or zero field strength in the vicinity of the transmitter. It is therefore necessary to 'fill' these nulls (i.e. to reduce their depth) so that reception in areas near the transmitter is not degraded. The nulls can be filled by shifting the phase of the currents in one or more of the tiers in the array so that the minimum field strength in the sidelobe region does not fall below about 10% of the maximum. Null filling results in a reduction of gain in the direction of the maximum of the main lobe and it is therefore important that the minimum amount of phase shift necessary to achieve acceptable reception is used. The loss of gain for '10% null filling' is about 0.4 dBD.

The beamwidth of the main lobe decreases as the aperture of the antenna increases and is given approximately by $57/A$ degrees where A is the aperture in wavelengths. For example, a 600-MHz antenna with a 20-wavelength (10-m) aperture will have a beamwidth of 57/20 = 3° (approximately). As will be seen from Figure 4.13, with a narrow horizontal beam, the signals in the service area will be received from the part of the radiation pattern that is below the

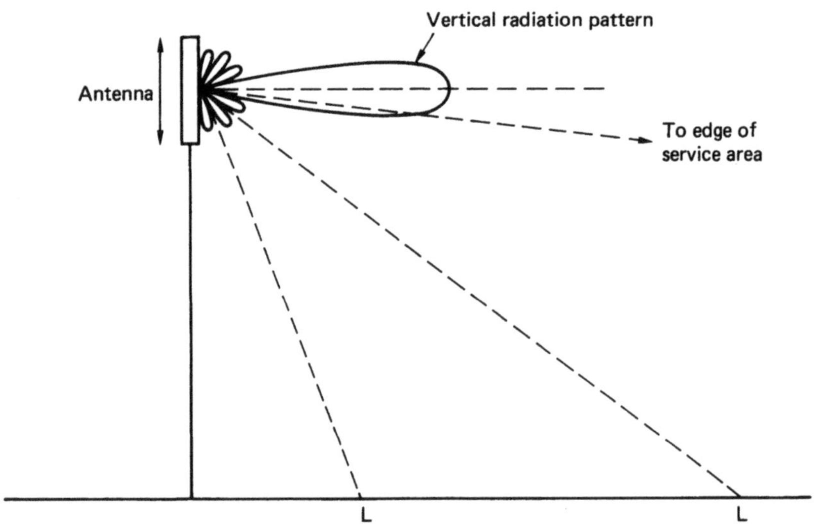

Figure 4.13 Nulls in vertical radiation pattern. L, areas of poor reception due to lack of null filling and beam tilt

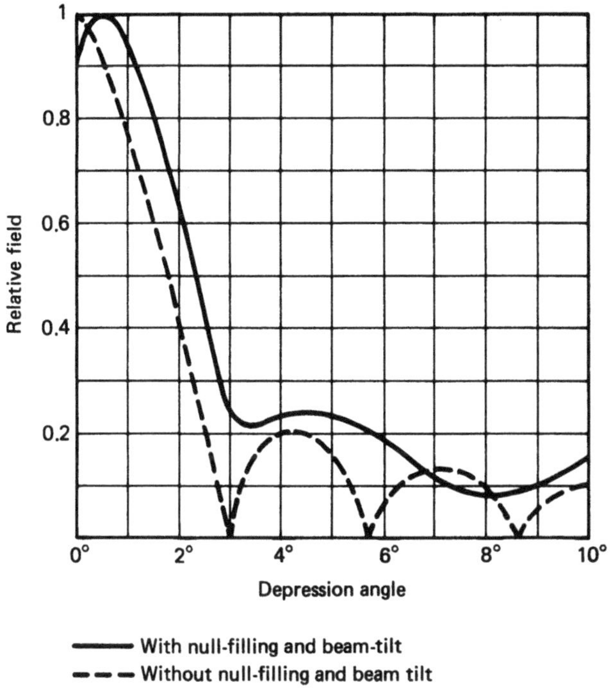

Figure 4.14 Eight-stack UHF antenna with null filling and beam tilt

maximum and this, of course, results in a loss of gain and reduction of field strength in the service area. For this reason, it is common practice to 'tilt' the beam below the horizontal to achieve optimum coverage of the service area. The beam-tilt depends on the main lobe beamwidth and the extent of the service area and, in practice, is in the range 0.5–2°. Beam-tilt is usually achieved by progressively changing the phases of the currents in the tiers of the antenna, and a typical example of a VRP with null filling and beam tilt is shown in Figure 4.14.

(c) Horizontal radiation pattern

In the majority of applications where the transmitting station is sited in the centre of the service area an omnidirectional radiation pattern is required. However, it is sometimes necessary to site the transmitting station at, or towards, the edge of the service area and, in this case, a directional radiation pattern is required. The panel type of antenna is particularly suitable for the generation of a directional radiation pattern by omitting some of the radiating elements on one or more faces of the mast or by changing their amplitudes and/or phases. VHF and UHF antennas of the pre-assembled 'cantilever' type can be designed to provide either an omnidirectional or directional horizontal radiation pattern and are usually cheaper than the panel type. However, in the case of a cantilever directional antenna, there is less flexibility at VHF as regards the type of horizontal radiation pattern that can be achieved than is the case with a panel antenna. Further, when antennas have to be erected one above the other on a common mast (a 'collinear' arrangement) the panel antenna has obvious advantages.

(d) Impedance

If the antenna impedance is not matched to that of the feeder, energy will be reflected back down the feeder to the transmitter. In general, the impedance 'looking back into the transmitter' will not match the feeder impedance and energy will therefore be re-reflected back up the feeder and radiated by the antenna as illustrated in Figure 4.15. This radiated, reflected energy is a delayed version of the transmitted signal and its effect depends on whether the

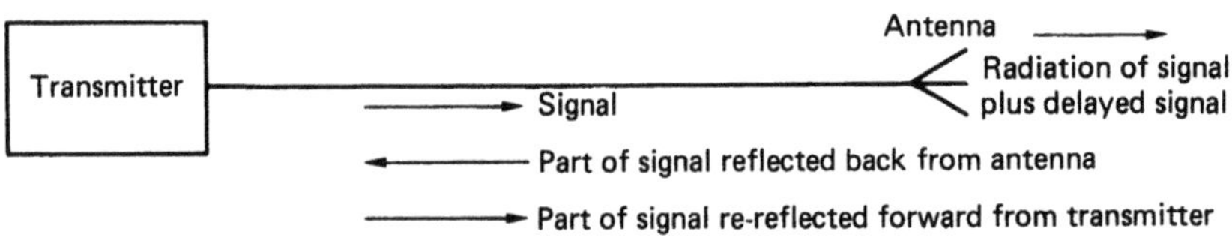

Figure 4.15 Effect of antenna mismatch

transmission is VHF/FM or television. In the case of VHF/FM, the delayed signal will result in degraded reception due to 'multi-path' distortion. In the case of television, the delayed signal will appear on the viewer's screen as a delayed image. Experiments have shown that the amplitudes of the delayed signals radiated by television (vision carrier frequency) and VHF/FM radio antennas mounted on tall structures (i.e. using long feeders) must not exceed 1.25% and 15% respectively if they are not to be apparent to the viewer or listener.

(e) Bandwidth and power-carrying capacity

The bandwidth of an antenna is the frequency range over which the performance as regards impedance (the impedance bandwidth) and radiation pattern (the radiation pattern bandwidth) remains within specified limits. In general, the change of radiation pattern with frequency of an antenna with a branch feeder system (see Appendix 2) does not limit the bandwidth and a radiation pattern bandwidth in excess of 2:1 is not difficult to achieve. With careful design, and the incorporation of matching facilities in the radiating

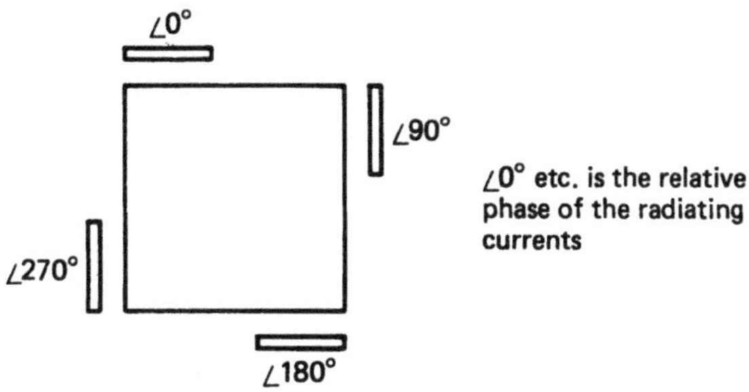

Figure 4.16 Offset panels on square mast to achieve impedance/bandwidth by reflection cancellation

elements and branch feeder system, it is possible to achieve an impedance bandwidth approaching 2 : 1. As a result a single antenna can be designed with a bandwidth covering the whole of a broadcast band. Thus Band I, Band II, Band III and Bands IV and V can each be covered by a single antenna. To give a wide impedance bandwidth with an omnidirectional radiation pattern, some panel antennas use a 'quadrature phase' arrangement which results in the cancellation of the reflected energy from the radiating elements on a square cross-section mast. In this case, which is illustrated in Figure 4.16, it is necessary to offset the panels to achieve co-phased radiation in the diagonal directions where equal, co-phased fields are required to provide an omnidirectional radiation pattern. Another technique often used for CP VHF/FM antennas is to arrange for alternate tiers of panels to complement each other, and for this reason, if equality of coverage for several services from one antenna is required, it is prudent to always specify an antenna comprising at least two tiers of panels.

Particularly in Western industrialized countries, it is often necessary to radiate more than one programme from a transmitting site and the wide bandwidth of VHF and UHF antennas makes it possible to radiate a number of transmissions from a common antenna. The number of transmissions that can be radiated is limited by the power-carrying capacity of the antenna and, in general, this depends on the number of radiating elements. For example, a basic UHF panel can have a power-carrying capacity of around 2 kW (mean), a high-gain UHF antenna will have around 64 panels, giving a power-carrying capacity of 120 kW, and a VHF antenna will have around 32 panels with a power-carrying capacity of 180 kW. The maximum transmitter power commonly used is around 60 kW in the UHF bands and 2 × 20 kW in the VHF bands so that, at these power levels, two UHF transmissions or five VHF/FM transmissions could be radiated from a single antenna, taking into account feeder and combining losses. At lower transmitter powers more than two transmissions can be radiated and UHF antennas radiating up to four transmissions are in use in many countries. In cases where the transmitter power is too high to enable transmissions to be radiated from a single antenna, separate antennas can be stacked above each other on the same mast.

4.3.4 Reliability, monitoring and maintenance philosophy

Well-designed UHF and particularly VHF antennas are very reliable, but when a fault does occur it may take a long time to repair. The antenna may be inaccessible for long periods due to icing, or a main feeder fault may require a new feeder, which is sometimes only available on long delivery from a manufacturer. Radiation hazards or the method of access to or through antennas may make it necessary to shut down a 'good' antenna in order to carry out repairs on a faulty one. As mentioned previously (Section 4.3.1) the 'split' antenna arrangement has much to offer and the overall antenna–transmitter

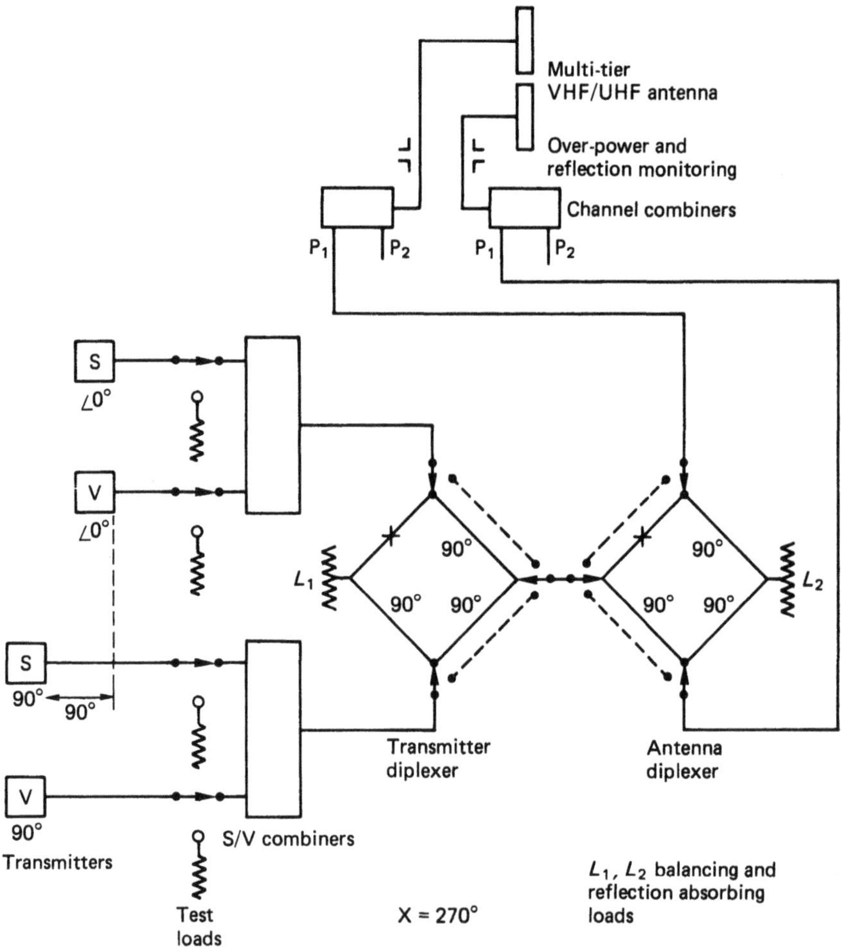

Figure 4.17 High-reliability high-power television parallel transmitter/split antenna system (comprehensive diplexer by-pass switching shown dotted)

system reliability is very high if such antennas are fed by pairs of transmitters operating in parallel (though, of course, this will be of little avail if the mains power for the transmitters and the programme feeds do not have comparable reliabilities). With this arrangement a single antenna or transmitter fault will reduce the ERP to the majority of the service area by only 6 dB, and this, power rating permitting, can be reduced to 3 dB at a convenient interval in programme transmission if the necessary switching is provided. However, the 'split' antenna system does have the disadvantage that halves of high gain antennas operated independently have null patterns very different from that of the complete antenna and consequently give rise to large differences in field strengths near to the transmitter. For this reason, when antennas of this type are sited in populated areas, it is unwise to use designs with excessive gain. Figure 4.17 shows a schematic arrangement of a split antenna system. If the arrangement is used for television, and areas in the directions of the nulls in the antenna pattern are to be served, it is important to first combine the outputs of the transmitters and then divide the power to feed the two halves of the antenna. The necessity for doing this arises because the transmissions radiated in the directions of the nulls are the small difference signals between the signals radiated by the two halves of the antenna and any inevitable small difference in the modulation linearity of the transmitters would be greatly magnified and seriously distort the grey scale of received pictures unless the outputs were first combined.

At VHF, and more frequently at UHF, it is sometimes difficult to design (because of the ratings of available feeders and lack of space) antennas with reasonable safety factors. The result is that a relatively minor fault or adverse weather conditions (causing standing waves) may endanger the whole of the antenna and feeder system. For these reasons, on such antenna systems it is prudent to incorporate over-power and reflected-power monitoring devices to trip the transmitter whenever there is a risk of catastrophic failure.

Experience has shown that the frequent dismantling of antenna systems for inspection, with a view to minimizing the risk of failure, actually increases the risk of failure. Instead it is recommended that antenna systems are visually inspected and checked for hot spots and reflected power levels monitored at regular intervals, and only if incipient faults are indicated is the system dismantled.

Figure 4.18 A 205-m tower sited in an urban area and used to radiate four UHF television programmes, each of 1 MW ERP, VHF radio services and carry communication and microwave link antennas. The television programmes are radiated from two relatively small aperture two-channel cantilever antennas inside a GRP shrouding cylinder each fed via duplicate feeders with pairs of 40-kW vision transmitters per programme. Unusually, but because this transmitting station and its dependent relay stations serve about 13 million people, reserve panel antennas are mounted on the tower column immediately below the main antennas (Courtesy of Engineering Information, BBC, London)

Figure 4.19 View inside a high-power UHF antenna showing distribution feeders and restricted space (Courtesy of Alan Dick & Co, Cheltenham)

Chapter 5

Dedicated microwave, UHF, VHF, HF and commercial programme-link systems

5.1 Introduction

The siting of a transmitting station is determined by the need to cover the required service area within the constraints of available services (power, water and access). As a result the optimum transmitting site may be remote from the programme source and a programme feed must be provided. If the station lies within a few kilometres of a main telecommunications route the provision of a programme feed may not be difficult. However, in many cases, and particularly in the Third World, the transmitting site is remote from a telecommunications route and the extension of the nearest route to the site would be prohibitively expensive. In this case a programme link is necessary to provide a programme feed and a number of alternatives are available:

1. Microwave links operate in the microwave band which extends from 1.7 to 11.7 GHz (1700–11 700 MHz). They provide a very wide modulation bandwidth which makes them particularly suitable for television applications, when the sound signal is often transmitted on a sub-carrier added to the video signal before modulation of the carrier. The links are normally designed for line-of-sight (i.e. unobstructed) operation and with proper design a very high quality and reliability can be achieved over distances up to about 50 km (30 miles). The disadvantage is that they are relatively expensive.
2. VHF and UHF links operate in the band which extends from 150 to 960 MHz. When compared with microwave links they have the advantages that, due to the lower frequency, they can be operated (to a degree decreasing with increasing frequency) over obstructed paths and are cheaper. However, because of the use of a lower frequency band, the modulation bandwidth is less than that of a microwave link and, in general, VHF and UHF links are used for high-quality monaural and stereophonic studio-to-transmitter applications.
3. HF links operate in the band 6–26 MHz and are used for monaural radio programme links. They have the advantage that very long distances (up to several thousand kilometres) can be covered but, due to the use of ionospheric transmission, the reliability and received programme quality

are relatively low. As a result, HF links are used only as 'feeders' for HF relay stations, which themselves use ionospheric transmission with the attendant low programme quality. They are, however, tending to be replaced by satellite links which offer higher quality feeds, but at higher costs.

4. Re-broadcast Reception (RBR) links are used to provide a programme source for one station by direct reception of another. They are mainly used in the VHF and UHF bands to feed individual stations or short chains of relay stations. They are, however, unsuitable as a means of distributing programmes to a whole network of transmitters as, in general, they cannot be engineered to provide acceptable quality when many links are used in series. However, it often makes economic sense to use them initially, replacing them with higher quality but more expensive microwave links as transmitter networks are extended.

5.2 Link fundamentals

The reliability and performance of any link will be dependent on climatic variations over the link path, the reliability of the equipment itself and the power supply to operate it. No system can be made to be 100% reliable, but some links can be engineered to approach this reliability. When designing a link system the first step is to decide on the percentage of time when the quality of the received signal must not be less than defined standards – clearly the performance of a link to a main station serving several million people should logically be higher than any RBR links it provides to small dependent relay stations. In general, taken together with the noise figure of the receiver, this will determine the minimum acceptable receiver input power, and it is thus possible to determine the required link transmitter power, taking into account the various losses over the transmission path and the gains of the transmitting and receiving antennas.

Assessment of the performance of a programme link commences with the calculation of the free-space loss, which is the loss of signal that would occur if the link were operated remote from the earth. To this loss must then be added the losses that will occur due to the nature of the actual path and these vary with the frequency band. In the VHF, UHF and SHF bands, the path losses depend on the terrain and are the refraction, diffraction, reflection and clutter losses. In the case of HF links, the path losses are the ionospheric and earth reflection losses as discussed in Chapter 3.

5.2.1 Free-space loss

In all bands the free-space loss of a link is given by a well-known formula due to Friis. This loss is the ratio of the power delivered by the transmitter and the power delivered to the receiver and is given by:

$$\text{Loss (dB)} = 32.4 + 20\log f + 20\log d - G_T - G_R$$

where f is the frequency in MHz, d is the path length in km, G_T is the transmit antenna gain in dBi, and G_R is the receive antenna gain in dBi. For example, at 800 MHz over a path length of 60 km and with transmitting and receiving antenna gains, both equal to 10 dB, the free-space loss will be 106 dB.

5.2.2 Refraction and equivalent earth radius

The density of the atmosphere decreases with increasing height above ground and, as a result, the transmitted signal follows a curved (refracted) path as illustrated in Figure 5.1(a). However, it is possible, as shown in Figure 5.1(b), to represent the signal path as a straight line if the radius of the earth is increased by

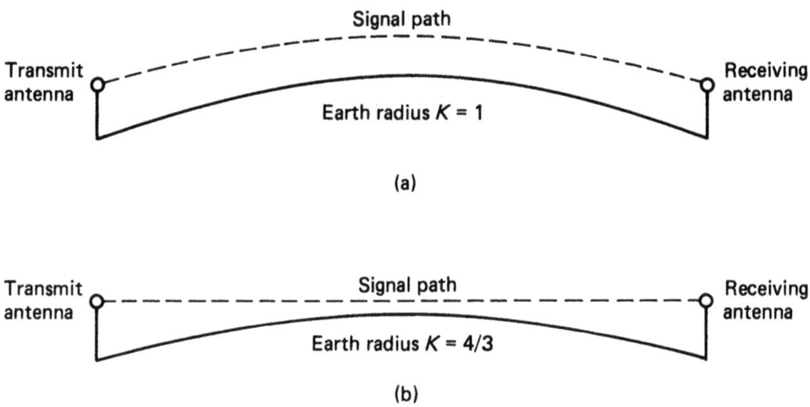

Figure 5.1 Earth radius factor. (a) Actual earth radius and refracted (curved) signal path. (b) 4/3 earth radius and unrefracted (straight) signal path

a factor K. The factor K, known as the 'earth radius factor', depends on the temperature and climate and can vary between infinity (i.e. the earth is flat) and about 2/3 under extreme weather conditions but the value obtaining under normal conditions is usually taken as 4/3 (1.333).

5.2.3 Diffraction loss

Diffraction loss occurs due to the presence of obstacles in the signal path and has been discussed in Chapter 4 and illustrated in Figures 4.4, 4.5 and 4.6. Diffraction loss increases with frequency and it is useful to remember that, at all frequencies, it is equal to 6 dB at grazing incidence.

5.2.4 Reflection loss

As illustrated in Figure 4.1, the signal arrives at the receiving antenna via two paths (directly and after reflection by the earth). The earth-reflected signal is reversed in phase and, where the difference in direct and reflected path lengths is small (as it normally is), the two signals tend to cancel. The amplitude of the reflected signal clearly affects the reflection loss and depends on the reflection coefficient of the ground. The reflection coefficient increases as the angle of arrival of the signal at the ground is decreased. Thus the reflected signal is greatest at grazing incidence, and ground near the transmitting antenna has little effect on its magnitude.

5.2.5 Clutter loss

Obstructions in the vicinity of the transmitting and receiving antennas will result in a reduction of the received signal known as 'clutter loss' and this increases with increasing frequency. Although clutter loss is an important factor in the calculation of broadcast coverage (particularly in the UHF bands), it is of less importance in the case of microwave links. This is because the transmitting and receiving sites for point-to-point microwave links are chosen to be as free as possible from local obstructions.

5.3 SHF line-of-sight links

SHF links take advantage of the very short wavelength which enables very high gains to be achieved with physically small transmitting and receiving antennas. As a consequence, low transmitter powers can be used and, in typical cases, the power seldom exceeds 5 W.

5.3.1 Antennas

Virtually all SHF antennas are of the parabolic dish type but the method of feeding varies and four types are shown in Figure 5.2. The most commonly used type is the 'front-fed' type but although it is the cheapest the fact that the feed system lies in the path of the transmitted signal results in 2–3 dB loss of gain. In

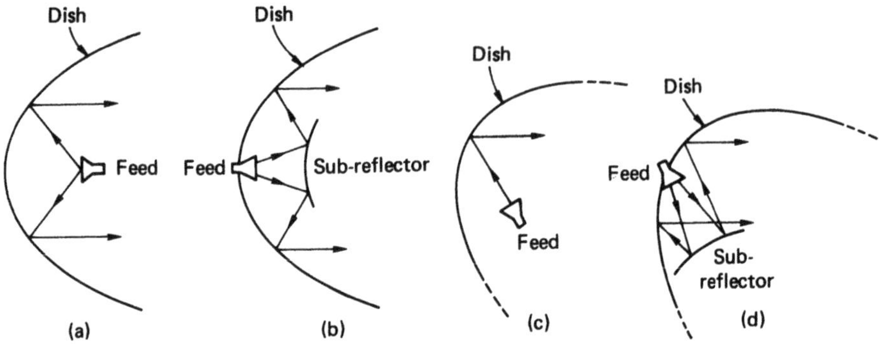

Figure 5.2 Methods of feeding SHF dishes. (a) Front-fed type. (b) Cassegrain type. (c) Front-fed offset type. (d) Cassegrain offset type

cases where maximum gain is vital (e.g. for satellite links) the Cassegrain, front-fed offset or Cassegrain offset type can be used. In the Cassegrain type the sub-reflector is arranged to be transparent to the transmitted signal and in the other two types the feed arrangement is kept clear of the path of the transmitted signal.

The gain of the front-fed type of parabolic dish commonly used for SHF programme links is given by:

$$G = 17.4 + 20 \log (Df) \text{ dBi}$$

where G is the gain, D is the dish diameter in m, and f is the frequency in GHz. For example, a 3-m-diameter dish antenna with an operating frequency of 7 GHz will have a gain of 43 dBi.

When calculating path loss it is necessary to take account of feeder loss, which will depend on frequency and feeder length. Except, occasionally, at 2 GHz (the bottom of the SHF band), waveguides are used as feeders from the transmitter or receiver to the antenna and the losses are approximately as shown in Table 5.1.

Table 5.1

SHF band (GHz)	*Feeder type*	*Loss (dB/100 m)*
2	1–1/4 in coaxial	5
4	Waveguide*	2.8
6	Waveguide*	4.6
7/8	Waveguide*	8.9
11	Waveguide*	11.5
12/13	Waveguide*	14.8

*Rigid-rectangular or elliptical.

The increase in feeder loss with frequency is offset by the increase in antenna gain that is possible with increasing frequency. For example, the effective gain (gain minus feeder loss) for a 3-m-diameter dish with a 50-m feeder length will be shown as in Table 5.2.

Table 5.2

SHF band (GHz)	*Antenna gain (dBi)*	*Feeder loss (dB)*	*Effective gain (dBi)*
4	39	1.4	36.7
6	43	2.3	40.7
7/8	45	4.5	40.5
12/13	49	7.4	41.6

There are three types of waveguide: rigid-rectangular, semi-flexible-elliptical and circular. The circular waveguide has the lowest loss but has the disadvantages that only virtually straight runs can be used and complex terminating networks are required. The losses of the rigid-rectangular and elliptical types are similar and the latter, which can be installed in a single continuous length, is generally cheaper. However, the elliptical waveguide needs to be installed with great care since small dents and deformations can result in mismatch and severe echo-distortion noise.

Dish antennas impose high wind loadings and, unless a stayed mast was specifically designed for the purpose, it is prudent to plan to mount dishes near to stay attachment levels as such arrangements are more likely to be structurally acceptable.

5.3.2 Path profiles and Fresnel clearance

Obstruction losses in the SHF band are high and, to enable a link to operate over distances up to about 50 km, a 'line-of-sight' signal path must be achieved. This means that a straight line drawn between the transmitting and receiving antennas, taking account of the chosen equivalent earth radius, must be clear of all obstructions. To ensure that there is no obstruction loss there must be a minimum clearance equal to the 'first Fresnel zone', which is an ellipse containing every point for which the sum of the reflected path distances to the two ends of the path is exactly one-half wavelength longer than the direct end-to-end path as illustrated in Figure 5.3. As will be seen, the Fresnel zone clearance reaches a maximum halfway along the path and its magnitude depends on the length of the path. The Fresnel clearance required over an obstacle is given by:

$$F_1 = 17.3(ab/fd)^{1/2} \text{ metres}$$

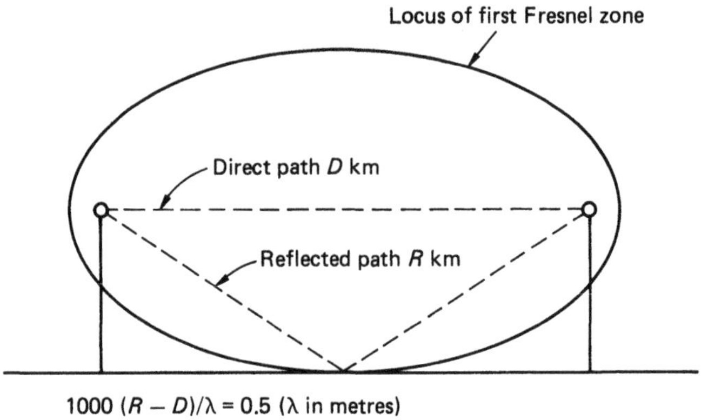

Figure 5.3 First Fresnel zone

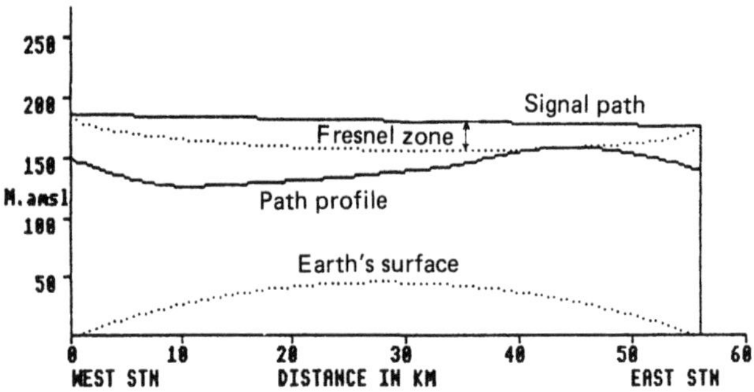

Path profile for path F3

Height of west mast:	37 M
Height of east mast:	37 M
Fresnel zone clearance:	1F1
Effective earth radius:	1.333

Figure 5.4 Link path with Fresnel zone clearance

where a is the distance of the obstacle from the transmitting terminal in km, b is the distance of the obstacle from the receiving terminal in km, f is the frequency in GHz, and d is the path length in km (i.e. $a+b$). For example, at 6 GHz the minimum (mid-path) clearance for a 40-km path length is 55 m.

Computer programs are available that will carry out the computations necessary to determine the antenna heights that will provide minimum first Fresnel zone clearance of the signal path over obstructions at all points along the path. In practice, a limited number of terrain heights and their corresponding distances from the transmitting terminal will be available and these are stored in a separate file for each path. The number of stored terrain heights depends on the length of the path and the scale of the map available and is, typically, between 10 and 30. It is therefore necessary for the computer program to reconstruct the path. To this end the computer program includes an interpolation routine that generates a large number (typically 100) of equally spaced terrain heights along the path irrespective of the number of stored heights. This provides the best available approximation to the actual profile but the accuracy obviously depends on the number of terrain heights in the file. The computer program then determines the mast heights necessary to achieve a minimum of at least first Fresnel zone clearance at all points along the path. Figure 5.4 is a typical example of a computer printout of an interpolated path showing the line-of-sight signal path between the transmitting and receiving antennas, the required antenna heights and the first Fresnel zone clearance along the path.

5.3.3 Reliability and outage time

With any point-to-point microwave link, fading will occur and will result in lost or degraded reception. The fading results from changes in the refractive index of the air between the transmitting and receiving antennas which cause the microwave beam to be tilted up or down and thus reduce the field strength at the receiving antenna. Fading is most likely to occur on hot, still, humid wind-free nights but can also occur in the daytime. Fading can also occur due to reflection of the beam from the ground, particularly when a lake or smooth, flat surface exists at some point along the path or if the stability of the structure carrying the antennas during high wind is inadequate. The reliability of a link is defined as the percentage of time that the signal at the receiver input terminals is above the minimum that will give acceptable reception. The parameter used to calculate the loss of reliability due to fading is the difference in receiver input power under non-fading conditions and the acceptable minimum input power, which is known as the 'fade margin'. A typical figure for the acceptable minimum receiver input power is around −70 dBm. Summarizing, therefore, the reliability of a link depends on:

1. The frequency
2. The length of the path
3. The fade margin
4. The type of terrain
5. The climate

Table 5.3 illustrates the effect of these factors for the typical case of a frequency of 6 GHz (SHF mid-band), a path length of 50 km, a fade margin of 40 dB (a typical value) and for a range of climates and terrain types.

Table 5.3

Terrain	*Climate*	*Reliability (%)*	*Outage times* (min.)*
Smooth	Hot/humid	99.9890	58
	Temperate	99.9945	29
	Dry	99.9972	15
Average	Hot/humid	99.9972	15
	Temperate	99.9986	7
	Dry	99.9993	4
Rough/rocky	Hot/humid	99.9993	4
	Temperate	99.9997	2
	Dry	99.9998	1

*Number of minutes per annum.

Even the lowest reliability figures in Table 5.3 appear, at first sight, to be quite high but the annual outage times show that this is not so. For example, a break of 58 minutes would be disturbing to listeners or viewers whether it occurred as a single break or as a large number of short breaks. As a result, for broadcasting it is desirable to aim at a minimum reliability of 99.999% (a maximum annual outage time of about five minutes). On long and difficult paths, particularly at high frequencies, it may not be possible to achieve this level of reliability. Further, on very long paths it may be necessary to use two or more links in tandem and the reliability will be reduced (for example, two links each with an outage time of 30 minutes will have an overall annual outage time of one hour). In these cases the reliability may be improved by space-diversity reception where two receiving dish antennas are mounted on the mast at heights separated by 10–15 m. The post-detection signals from the two dishes are either combined or that with the maximum amplitude is selected. Space-diversity reception can give an improvement of up to 100: 1 with a 40-dB fade margin (for a link reliability of 99.99%, the overall reliability becomes 99.9999%, a reduction in annual outage time from 52 minutes to 32 seconds).

5.4 VHF and UHF links

VHF links operate in the bands 148–174 MHz, 215–240 MHz, 300–330 MHz, and UHF links in the bands 450–470 MHz and 890–960 MHz and are used mainly for relatively short-haul (up to 30 km) studio-to-transmitter radio programme circuits. They have the advantage that the link equipment, and particularly the antenna equipment, is cheaper than for SHF. Further, the use of lower frequency bands reduces the diffraction loss due to obstacles and the incidence of fading, so that UHF and, to a greater extent, VHF links can operate over paths obstructed for SHF links. To avoid co-channel interference with other services, relatively low powers (up to 15 W) are normally used in conjunction with directional, relatively high gain antennas.

5.4.1 Planning considerations

A VHF or UHF link must provide an adequate signal-to-noise ratio (SNR) at the receiver and, as in the case of SHF links, there must be an adequate fade margin. For radio broadcasting purposes the fade margin is based on a receiver minimum input power level of around −67 dBm (100 μV/50 Ω) for monophonic broadcasting and −60 dBm (225 μV/50 Ω) for stereophonic broadcasting. For UHF television the corresponding input power level would be −47 dBm (1 mV/50 Ω).

In addition to the RF performance there are a number of practical factors that must be considered.

1. *Mechanical factors.* Loads imposed by winds, ice and snow may result in antenna failure and it should be remembered that aluminium and its alloys are prone to fatigue failure. Plastics do not corrode but can deteriorate due to the ultraviolet component of sunlight.
2. *Ice and snow.* Apart from imposing enhanced mechanical loads, ice and snow on an antenna can cause an increase in voltage-standing-wave ratio (VSWR) and loss of gain.
3. *Natural electrical effects.* Particularly in tropical areas, lightning storms can occur and can cause damage to antennas mounted at the top of masts or towers. The provision of a continuous, well-bonded path to earth will reduce the possibility of damage to the antenna. The transmitting and receiving equipment must be protected by good antenna design and system grounding.

   Precipitation static can cause severe interference at frequencies near the lower end of the VHF band when intense electric fields are generated in electrical storm conditions. The interference is caused by charged raindrops impinging on the antenna elements and can be eliminated by covering the receiving antenna elements with dielectric shields.

5.4.2 Antennas

Yagi antennas are very widely used for VHF and UHF links as they are cheap, simple to construct and provide a highly unidirectional radiation pattern. The Yagi is electrically simple as it comprises a single active (driven) element (i.e. only one element is connected to the feeder) together with a passive reflector and a number of passive directors as illustrated in Figure 5.5. Typical gain figures for a single Yagi antenna are given in Table 5.4.

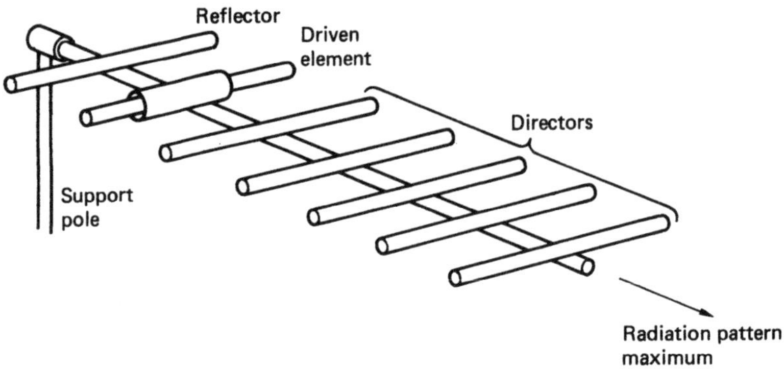

Figure 5.5 Yagi antenna

Table 5.4

Number of elements	*Gain (dBi*)*	*Gain (dBD*)*
3	7	5
4	9	7
6	11	9
8	13	11
12	15	13
15	16	14
18	17	15

*See definitions in Appendix 2.

Higher gains can be achieved if a number of Yagi antennas are arrayed as illustrated in Figure 5.6 for a four-Yagi arrangement. However, if it is structurally feasible, it is usually desirable to mount the Yagis in the same horizontal plane, as this increases the gain in the horizontal plane and this, in turn, minimizes the risk of co-channel interference. Provided the individual Yagis of an array are spaced correctly, the increase in gain (dB) is equal to $10 \log (N)$ where N is the number of Yagis. For example, from Table 5.4, an array of four eight-element Yagis would have a gain of $13+6=19$ dBi (17 dBD).

Where a wide antenna bandwidth is needed (e.g. if several radio transmissions are to be transmitted) log-periodic antennas (LPAs) are often used. The LPA is illustrated in Figure 3.7 and comprises a series of half-wave dipoles spaced along a transposed or dielectrically loaded transmission line. At any given frequency, only dipoles whose lengths are approximately a half-wave are excited and form an active region. Thus, at any frequency, the gain and impedance remain constant, resulting in a bandwidth limited only by the dimensions of the smallest

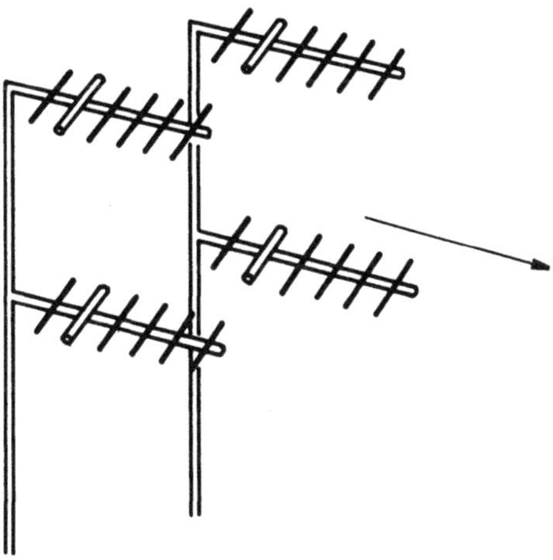

Figure 5.6 Array of four six-element Yagi antennas

and largest dipoles. The disadvantage of the LPA is that the gain is low, as only a few dipoles are used at any particular frequency. The gain of an LPA is, in fact, unlikely to exceed 9 dBi (7 dBD) and if higher gains are required it is necessary to use an array of LPAs.

For some applications, the 'grid dish' in which the paraboloid comprises close-spaced metal strips is used and has the advantages of robustness and simplicity, and a low wind load if not subject to icing. Further, a grid dish can be cheaper than an array of Yagis or LPAs. However, the gain of a dish of a given diameter depends on the frequency and, as a result, this type of antenna is not likely to be cost-effective except in the UHF band. For example, a 3-m-diameter dish will have a gain of 20 dBi at the lower end of the UHF band and 26 dBi at the upper end.

Table 5.5. Cable loss in dB/100 m

Frequency (MHz)	*Cable diameter (inches)*			
	$\frac{1}{4}$	$\frac{1}{2}$	$1\frac{1}{4}$	$1\frac{5}{8}$
100	4.3	2.1	1.0	0.7
200	6.1	3.1	1.4	1.0
300	7.6	3.8	1.8	1.3
400	8.8	4.4	2.1	1.6
500	10.0	5.0	2.4	1.8
800	12.8	6.5	3.1	2.4
1000	14.5	7.3	3.6	2.8

Figure 5.7 UHF relay station antennas: the transmitting antenna comprises eight crossed log-periodic antennas producing a three-lobed directional radiation pattern, whilst the receiving antenna comprises four Yagis giving a gain of over 20 dB/D (Courtesy of Engineering Information, BBC, London)

To obtain the effective gain of all these types of antennas it is, of course, necessary to deduct the feeder loss. VHF and UHF links use relatively low transmitter powers and the size of the feeder is governed by the need to keep the attenuation to an acceptably low level. Since power rating is normally unimportant, coaxial feeders with a foam polyethylene dielectric are often used. These feeders are highly flexible, easy to install and have the additional advantage that they do not need pressurizing to prevent the ingress of moisture. Table 5.5 shows the approximate attenuation of foam dielectric feeders of commonly used diameters.

5.4.3 Polarization

Both horizontal polarization (HP) and vertical polarization (VP) are used for VHF and UHF links. Since the transmitting and receiving antennas are never mounted at low heights above ground there is little difference so far as propagation is concerned. However, polarization can be chosen to minimize co-channel interference with other services. HP is usually the first choice since horizontally polarized dipole Yagis are naturally more directional in the horizontal plane, but different polarizations can be used for antennas with different services mounted on a common mast or tower. Circular polarization (CP), in which equal VP and HP fields are radiated in phase quadrature, is sometimes used as it reduces the effect of interference due to reflected multi-path signals even more than HP. These are delayed signals arriving at the receiving antenna after reflection by natural or man-made objects lying outside the direct path from the transmitting antenna.

5.4.4 Reliability

As with SHF links, fading due to changes in the refractive index of the air occurs and affects the reliability of reception. However, the fading is less severe than that encountered with SHF links. Unlike the SHF case, the dominant effect of ground reflection is not to give rise to additional fading but to cause a fairly constant reduction of the received field due to a cancellation of the direct wave by the ground-reflected wave rather than to contribute to the degree of fading. Thus, as in the case of VHF and UHF broadcast propagation described in Section 4.2.4, the effective heights of the transmitting and receiving antennas must be taken into account when calculating field strength.

Summarizing, therefore, the reliability of a VHF or UHF link depends on:

1. The frequency
2. The length of the path
3. The climate
4. The fade margin

The frequency and length of path will, of course, be fixed parameters for a particular path. The fade margin will depend on the design parameters of the link, which are:

1. The transmitter power
2. The transmitting and receiving antenna gains
3. The required minimum receiver power
4. The diffraction (obstacle) loss
5. The reflection loss

Tables 5.6 and 5.7 give the reliability for a range of path lengths and fade margins for VHF and UHF links under average climatic conditions.

Table 5.6. Reliability % – VHF

Path length (km)	*Fade margin (dB)*			
	10	20	30	40
10	99.9999	100.0000	100.0000	100.0000
20	99.9993	99.9999	100.0000	100.0000
30	99.9976	99.9998	100.0000	100.0000
40	99.9944	99.9994	99.9999	100.0000
50	99.9890	99.9989	99.9999	100.0000
60	99.9810	99.9981	99.9998	100.0000
70	99.9698	99.9970	99.9997	100.0000
80	99.9549	99.9955	99.9995	100.0000

Table 5.7. Reliability % – UHF

Path length (km)	*Fade margin (dB)*			
	10	20	30	40
10	99.9996	100.0000	100.0000	100.0000
20	99.9965	99.9996	100.0000	100.0000
30	99.9881	99.9988	99.9999	100.0000
40	99.9719	99.9972	99.9997	100.0000
50	99.9451	99.9945	99.9995	99.9999
60	99.9051	99.9905	99.9991	99.9999
70	99.7751	99.9849	99.9985	99.9998
80	99.7751	99.8775	99.9978	99.9998

Tables 5.6 and 5.7 show, clearly, that a low fade margin can be quite acceptable for a short path and that the required fade margin for a given reliability increases with path length and frequency. Suppose, for example, that a fairly high reliability of 99.999% (an annual outage time of about five minutes) is required. Tables 5.6 and 5.7 show that this can be achieved with the following fade margins:

1. VHF
   (a) A fade margin of 10 dB for paths up to 20 km
   (b) A fade margin of 20 dB for paths from 20 to 40 km
   (c) A fade margin of 30 dB for paths from 40 to 80 km

2. UHF
   (a) A fade margin of 10 dB for paths up to 10 km
   (b) A fade margin of 20 dB for paths from 10 to 30 km
   (c) A fade margin of 30 dB for paths from 30 to 60 km
   (d) A fade margin of 40 dB for paths from 60 to 80 km

To determine the reliability of a link it is necessary to determine the fade margin, and the first step is to calculate the free-space attenuation between the input terminals of the transmitting antenna and the output terminals of the receiving antenna, assuming that both antennas have a gain of 0 dBi. As in the case of SHF transmissions, this is given by:

$$\text{Loss} = 32.4 + 20 \log f + 20 \log d \text{ dB}$$

where f is the frequency in MHz, and d is the path length in km. Using the attenuation as a basis, the fade margin can then be determined from a knowledge of transmitted power, transmitting and receiving antenna gains, reflection loss and obstruction loss. The calculation of these losses is described in Section 4.2.4 and the link plannning process is best illustrated by an example.

A radio programme link is required over a path length of 20 km using a frequency of 300 MHz and it is required to find the antenna gains needed to give a reliability of 99.999%. From the above expression, the free-space attenuation will be 108 dB and the proposed link parameters are:

Transmitter power	15 W (42 dBm)
Reflection loss	18 dB
Obstacle loss	10 dB
Minimum receiver input power	−75 dBm

(For transmitter power, dB relative to 1 mW is given by 10 log (1000*W*) where *W* is in watts.)

For 0 dBi transmitting and receiving antenna gains, the power into the receiver is given by:

Transmitter power	+ 42 dBm
Free-space attenuation	− 108 dB
Reflection loss	− 18 dB
Obstacle loss	− 10 dB
Receiver input power	− 94 dBm

Table 5.6 shows that for a 20-km path at VHF a fade margin of 10 dB should give a 99.999% reliability. Since the minimum receiver input power has to be −75 dBm, the combined effective gain of the transmitting and receiving antennas must be 10 + 94 − 75 = 29 dBi, to give a fade margin of 10 dB. From Table 5.4 this gain could be achieved with 15-element transmitting and receiving Yagi antennas, each with a feeder loss of 1 dB. Longer paths than that considered in this example would obviously require an array of two or more Yagi antennas at each end of the link and/or diversity reception.

5.4.5 Re-broadcast reception (RBR) programme feeds

Frequently a convenient and economic method of solving the programme feed problem for a transmitting station is to receive and re-broadcast the programme from another transmitting station. Such an arrangement is often referred to as an RBR system. Clearly, the signal available from another transmitting station at normal domestic receiving antenna heights will not be at a high level since, if it were, listeners or viewers would have a service and there would be no need for the station or a feed. However, in many cases, an acceptable programme feed can be achieved by using sophisticated receiving equipment, a high-mounted receiving antenna of high gain and (sometimes) diversity reception. However, RBR is not a 'cure-all' for programme feed problems since there are fundamental limits to the range at which it can be successfully applied, and on-site conditions frequently reduce this further. The following paragraphs summarize the advantages, disadvantages and limitations of RBR systems in the different broadcast bands.

(a) MF

MF transmissions are seldom used to provide an RBR programme feed. This is because the dominant noise in this band is external to the antenna rather than generated in the receiver and cannot, therefore, be reduced by use of a sophisticated receiver. Further, the long wavelengths make it difficult to achieve high-gain directive receiving antennas and thus to improve on the SNR possible with simple antennas commonly used with domestic receivers. Also, the long wavelengths make it difficult to arrange for diversity reception to overcome fading because the two antennas need to be at least 20 wavelengths (several thousand metres) apart. Another difficulty is that, beyond the edge of an MF service area (where an RBR antenna would have to be sited), there is a fade zone where good reception is impossible. Beyond the fade zone, reception at night is by sky-wave and subject to fading which is particularly severe around dawn and dusk. However, in spite of its shortcomings, MF RBR has been used as a second class reserve feed for VHF radio transmitters in areas where a programme is duplicated on MF and VHF.

(b) VHF/UHF

The shorter wavelengths corresponding to frequencies in the VHF band (approximately 6 m in Band I, 3 m in Band II, 1.5 m in Band III, 0.6 m in Band IV and 0.4 m in Band V) make RBR programme feeds a practical proposition. Greatly improved reception of weak signals, when compared with normal domestic reception, is possible because:

1. High receiving antenna gains are possible (up to 20 dBi or more with arrays of multi-element Yagi antennas at UHF).

2. Receiving antennas can be mounted on the mast (at the station receiving the programme feed) at heights greater than is possible in a domestic installation. This can result in a substantial increase in field strength, particularly if tests are carried out to determine the optimum height for the receiving antenna.
3. Two antennas can be used, mounted at different heights, to take advantage of diversity reception. A spacing of 20 wavelengths or more is required and can be readily achieved. For example, at 100 MHz (in the FM radio broadcasting band) a spacing of 20 wavelengths will amount to about 60 m (197 ft).

VHF and UHF RBR links are usually used to feed VHF or UHF transmitting stations and, in general, the wider the receiving–transmitting frequency spacing the easier the system will be to engineer. But whatever the spacing, the approach to the problem is the same; can an adequate signal free from co-channel interference and other interference be received on the proposed transmitting site and, if so, could the level of interference from the locally transmitted programme(s) be reduced to an acceptable level, to permit the signal to be utilized?

The method of determining the first requirement in respect of receiver input power has been dealt with in some detail in Chapter 5, Sections 5.2 and 5.4 (if the parent station was already operational the signal would, of course, be measured rather than predicted) and the minimum signals for good monaural, stereophonic and television reception are quoted in Section 5.4.1 with further information in Sections A2.2 and A2.15 (Appendix 2). If co-channel interference or other interference is a problem it may be possible to reduce this to an acceptable level by using a phased or appropriately horizontally spaced array of Yagis to reject signals on specific bearings. Interference from the locally transmitted programme(s) can, in general, be reduced by spacing the receiving and transmitting antennas on the support structure as far apart as possible, and may be further reduced if the receiving antenna consists of two tiers of elements spaced by half a wavelength at the local transmitting frequency. However, filters tuned to reject the local transmissions are usually necessary in the feeder from the receiving antenna to the receiver, to prevent overloading, and band-pass filters may be required in the transmitting feeder to attenuate any radiated spurious frequencies falling in the channel being received, the latter being more likely to arise if solid-state, rather than valved output stage, transmitters are used. At UHF the coupling between the receiving and transmitting antennas is typically such that receiving input levels of −60 dBμ are acceptable for television with relatively simple filtering, provided the received and transmitted channels are separated by two blank channels. However, for VHF stereophonic services, where the coupling between antennas is greater, the minimum spacing is 0.4 MHz with a typical minimum workable input receiver power of −47 dBm.

At UHF, and more so at VHF, and particularly when high-power transmissions are involved, it is very important that there is good and constant contact between metallic connections. Any rubbing contacts, e.g. the pivots of stay connections, must be bridged by solid, but flexible, conductors. Contacts between dissimilar metals where corrosion is the likely consequence should be avoided. Failure to meet these conditions will result in arcing and the generation of wideband noise which is liable to cause unacceptable degradation of the incoming signal. Cases of this type of interference due to intermittent metallic contact in areas adjacent to sites have been observed, for example, between strands of rusty barbed wire. Interference originating from intermittent contacts should be suspected if it is observed to be reduced in wet weather – it seems that rain puts out the arcs!

When more than one high-power transmitter is operated on the same site, intermodulation products, which may fall on the received frequencies, will be generated unless the coupling between the transmitters is reduced to an adequately low level, by the use of filters and adequate screening in transmitter halls.

At sites using RBL feeds, alternative equipment which bring the signal down to base band or an intermediate frequency (IF) are available. The former are to be preferred for VHF radio services as they enable closer receive–transmit spacings to be handled. For UHF television equipment, using the IF principle is satisfactory.

Clearly, whilst VHF/UHF RBL links are attractive relative to microwave links on cost considerations, they are not without their problems and only rarely is it feasible to achieve the high quality associated with microwave links which is essential if many links are to be used in tandem. However, they are often used as reserve feeds for microwave links and when it is considered desirable to provide a service to a relatively small group of people when the available signal is less than that quoted in Section 5.4.1 and costs would rule out the provision of a microwave link. In such situations the local population may well prefer a service somewhat below the standards normally adopted for main transmitters to none at all, and, indeed, experience has shown this to be the case. Thus a main 625-line television transmitter may be engineered to radiate a service with an SNR of 44 dB, but lower ratios are capable of providing very acceptable pictures as the results of some subjective tests quoted in Table 5.8 shows.

Summarizing, picture quality worsens progressively with decreasing SNR, but much more rapidly for ratios below 31–32 dB. Thus it is seen that receiver input levels of −60 dBm can produce pictures of good entertainment value.

If, in practice, on a particular site, even an array of, say, four Yagis did not produce an acceptable level of signal, a low noise head amplifier adjacent to the Yagis may do so, but this is only recommended as a last resort as active devices on high structures are a maintenance liability.

There is, of course, a limit to the range at which a VHF RBR programme feed is feasible. This depends on the nature of the path (particularly the presence of natural obstructions), the ERP (effective radiated power) of the transmitting station, the transmitting antenna height and the type of transmission. Generally speaking, however, the following are typical maximum practical ranges for RBR links in the VHF/UHF bands:

1. 200 km for a VHF/FM monophonic or stereophonic transmission used to

Table 5.8

Measured signal-to-noise ratio	*Description of picture*
42	Threshold of excellence: better signal-to-noise ratios do not improve subjective appearance of picture
39	If it had not been possible to demonstrate pictures with a better signal-to-noise ratio, pictures would have been described as excellent
34	Noise does not detract from pleasure of viewing
29	Picture could obviously be better, but good entertainment value – quality unlikely to deter people from viewing. A community would almost certainly prefer a service of this standard to none at all
24	Viewer would be fully justified in complaining if a service of this standard was provided.

provide a programme feed for an MF transmitter or monophonic VHF/FM transmitter.

2. 120 km for a VHF/FM stereophonic transmission used to provide a programme feed for another VHF/FM stereophonic transmitter.
3. 120 km for a VHF colour television transmission used to provide a programme feed for another VHF or UHF colour television transmitter.
4. 100 km for a UHF colour television transmission used to provide a programme feed for another VHF or UHF colour television transmitter.

5.5 HF radio links

5.5.1 Planning objectives

As described in Chapter 3, HF transmitting stations are used either for long-distance external broadcasting (i.e. broadcasting to other countries) or for wide-area domestic broadcasting within a country. For domestic broadcasting the transmitting station location is not usually critical because a country can be covered from sites spread over quite a wide area. Thus, for domestic broadcasting, the transmitting station can be sited adjacent to a telephone route or adjacent to the programme-originating studios and there is normally no requirement for a programme link. In the case of external broadcasting, as with domestic broadcasting, there is usually no difficulty in siting a station adjacent to the programme source but there can be a coverage problem. This arises because the effective coverage range of a long-distance HF transmission is limited to about 10 000 km by attenuation and fading. Thus, if it is necessary to extend the service to distances further than 10 000 km, a 'relay station' must be provided. Here it is worth pointing out that, unlike relay stations in other bands, which usually have low powers, an HF relay station will usually have a number of high-power transmitters. For example, the BBC World Service is extended to south-east Asia by means of a high-power, multi-transmitter relay station sited at Singapore.

It is now possible to provide a high-quality programme feed to anywhere in the world by satellite transmission but this requires an expensive ground station and in some cases an HF programme link is therefore used. Also, HF programme links are used to provide a 'standby' in the event of a failure of the satellite link. As has been made clear, unlike VHF, UHF and SHF links, which have a range up to 200 km at most, HF links can be operated over paths several thousands of kilometres in length. However, an HF link has a limited bandwidth and, because the transmission is subject to ionospheric variations, has a relatively low reliability and is most suited to speech and light entertainment programmes rather than the full range of programmes used in domestic broadcasting. As a result, HF links are almost invariably used for transmissions that are to be re-broadcast at HF.

5.5.2 Reliability

As noted in Section 3.2.3, reliability in the HF band is defined in the same way as in the higher frequency bands and is the percentage of time during which the received SNR exceeds a specified value. However, performance standards are lower in the HF band and, as a result, lower reliability figures are acceptable. A typically acceptable reliability figure for an HF link is 95% but this does not mean that there is no usable received signal for 5% of the transmission time. The nature of HF fading is such that 95% reliability can mean that the link is unusable for less than 1% of the time and gives sub-standard performance for about 4% of the time.

5.5.3 Transmitters

Single-sideband (SSB) transmission (rather than double-sideband (DSB) transmission) is used for HF links and it offers important advantages:

1. The SSB transmitter power required is less than that for a DSB transmission. For example, with similar transmitting and receiving antenna gains, a 50-kW PEP (peak-effective-power) SSB transmitter will give the same performance as a 100-kW DSB transmitter. Further, an SSB transmitter transmits only one sideband and does not transmit a carrier and, as a result,

the power consumption is only about a tenth of that of a DSB transmitter, giving the same performance.

2. With DSB transmissions, selective fading can occur when the two sidebands suffer different ionospheric phase shifts during transmission and cancel at some audio frequencies. With SSB transmission this distortion is obviated.
3. The bandwidth of an SSB transmission is only half that of a DSB transmission and thus increases the number of channels available and reduces the incidence of co- and adjacent-channel interference.

As a result, SSB transmitters for HF link applications typically have a maximum output power of 50-kW PEP (as opposed to the 500-kW maximum output power of DSB broadcast transmitters) and because of the low mains power requirement are very economical to operate.

5.5.4 Antennas

The required performance of transmitting and receiving HF antennas as regards bandwidth, gain and TOA (take-off angle) depends on the path length. To illustrate typical requirements, Table 5.9 shows the mode (F1 means one-hop via the ionosphere, etc.), the TOA and range of frequencies required for a range of path lengths.

Table 5.9

Path length (km)	*Mode*	*TOA (degrees)*	*Frequency range* (MHz)*
200	F1	71	3–8
500	F1	48	3–10
1000	F1	28	5–14
2000	F1	12	7–21
4000	F2	12	7–21
5000	F2	8	8–23

*The range of transmitted frequencies required to maintain communication over the diurnal, seasonal and sunspot cycles.

It will be seen that the transmitting and receiving antennas need to cover a frequency band of more than 2:1 in all cases and a constant TOA must be maintained over this bandwidth. The most suitable type of antenna meeting this requirement is the LPA, the principles of which are described in Chapter 3. Further, by mounting an LPA as illustrated in Figure 3.8(a), it is possible to maintain a constant electrical height above ground and, consequently, to maintain a constant TOA. Figure 5.8 shows examples of LPAs suitable for use as HF link transmitting antennas.

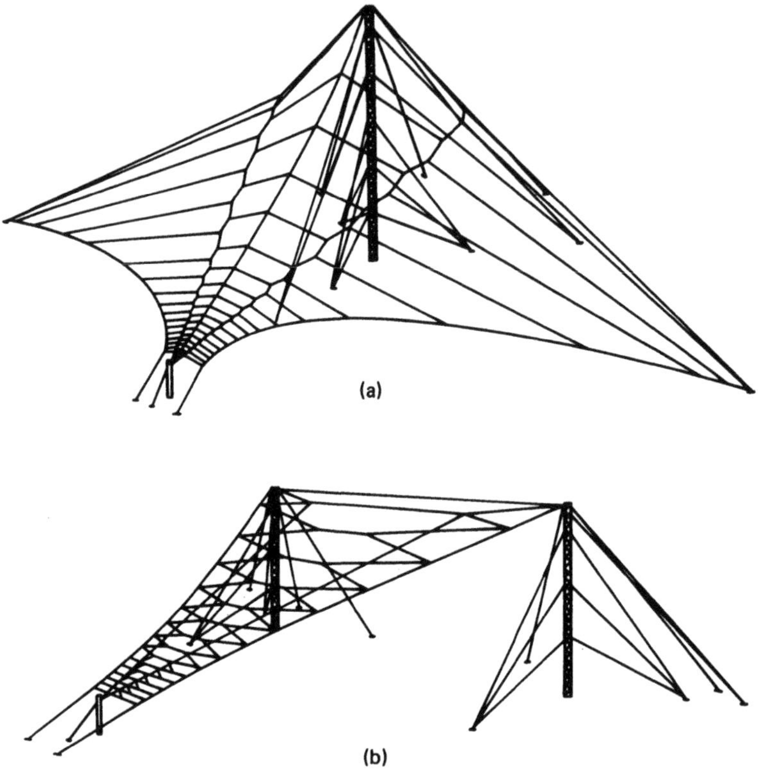

Figure 5.8 HF log-periodic antennas. (a) Medium range: TOA, 30–40°; gain, 9–14 dBi. (b) Long range: TOA, 16°; gain, 16.5 dBi (Courtesy TCI)

Table 5.10

Path length (km)	*Transmitter power (kW PEP)**	*Antenna gain (dBi)*	
		Transmitting	*Receiving*†
200	5	7	7
500	5	8	8
1000	10	9	9
2000	20	10	10
4000	40	14	14
5000	40	15	15

*The transmitted powers assume SSB transmission.
†The receiving antenna gain is in fact the directive or intrinsic gain (the gain neglecting losses) since incoming noise determines SNR in the HF band.

Figure 5.9 Loop antenna receiving array (Courtesy TCI)

So far as receiving antennas are concerned, it is possible to use an antenna designed for transmitting applications and many HF links use such antennas. However, since the received SNR at HF is determined by incoming noise it is possible to use a much less efficient and cheaper antenna. A practical antenna of this type is a small, vertical, balanced loop about 1.5 m in diameter and mounted about 2 m above ground. A number of loops can be arrayed to provide a wide variety of directional patterns and a typical loop antenna array is shown in Figure 5.9.

To give an indication of the transmitter powers and antenna gains required in practical applications, Table 5.10 shows typical values as a function of path length to achieve a reliability of 95%. The transmitting and receiving antenna gains are assumed to be equal but, obviously, one can be decreased and the other increased by the same amount to achieve the same reliability.

5.6 Nationally or commercially owned communication networks as programme feeds

Where communication networks are already established, their exclusive use, or use in combination with a 'broadcasters' equipment, may be the most economic way of linking a studio centre to a network of transmitters. Increasingly, such networks are employing digital techniques to transmit information and charges are based on the capacity or 'bit-rate' (equivalent to bandwidth). Thus techniques which reduce the bit-rate or bandwidth required for programmes are particularly desirable with this type of charging. The BBC has developed equipment to do this, which is manufactured under licence. In the case of television – called 'Sound in Syncs' – the audio signal – mono or stereo – is coded digitally and inserted into the line-synchronizing pulses of the video waveform prior to distribution to transmitters. At the transmitters it is removed and used to modulate the FM sound carrier. For radio services a system called NICAM – near instantaneous companded audio multiplex – is used. In this system the audio signals are coded digitally with 13 bits per sample of which only the 10 most significant are selected for transmission. The system takes advantage of the fact that higher levels of quantizing noise are found to be acceptable during loud parts of programmes.

Communication networks of this type are generally of high quality and reliability. They may use coaxial cables, but are more likely to use microwave links or fibre optic cables.

Chapter 6

LF, MF and HF radio transmission

6.1 Introduction

Radio broadcasting commenced with the introduction of LF and MF domestic transmission in the 1920s, which was followed by a long-distance HF transmission in the 1930s. Over the years, there have been many important developments in the design and performance of LF, MF and HF transmitter equipment and it might be thought that there would be little scope for further improvement. However, advances in solid-state circuits and in thermionic valve design have resulted in a continual increase in transmitter efficiency, reliability and output power together with a reduction in physical size. Further, the advances in solid-state techniques now make it possible to design cheap single sideband (SSB) receivers and hence to contemplate a change to SSB broadcasting in the HF band (and, possibly, in the LF and MF bands at some time in the future). It will be seen, therefore, that technical development in the field of radio broadcast transmission in the LF, MF and HF bands is far from over and will, without doubt, continue into the future.

6.2 Amplitude modulation systems

Transmissions in the LF, MF and HF bands use amplitude modulation (AM) in which the programme information is used to control (i.e. to modulate) the amplitude of a 'carrier' wave as shown in Figure 6.2(a). The 'modulation depth' (the amount by which the carrier amplitude is modulated) varies between 0% (the steady carrier) and 100% (the carrier amplitude varies between zero and twice its steady value). In the receiver the detector gives an output proportional to the 'envelope' of the received signal shown in Figure 6.2(a) so that the original programme information is recovered and can be amplified to the level required to drive a loudspeaker. Amplitude modulation results in an RF spectrum in which the programme information appears in the form of sidebands both above and below the carrier frequency as shown in Figure 6.2(b). This form of modulation is known as 'double-sideband (DSB) modulation'.

The task of the transmitter designer is to achieve the best possible solution as regards efficiency, compactness, ease of maintenance and minimum capital cost. An acceptable design is easier to achieve when a low output power is required, as the efficiency is then less important. However, for the higher output powers of

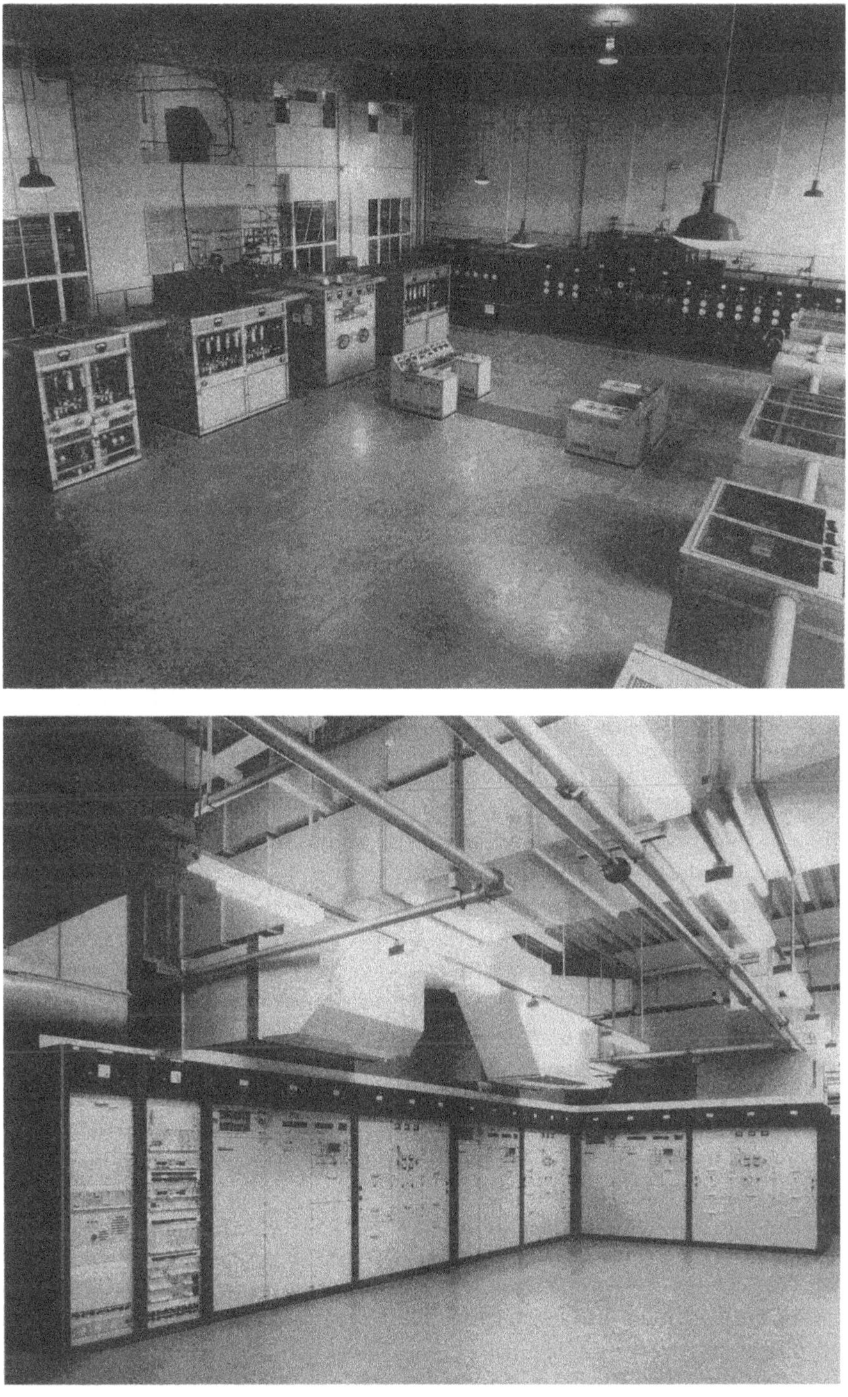

Figure 6.1 The original (1930s) MW transmitters (a) and their modern replacements (b) (Courtesy of GEC–Marconi Ltd)

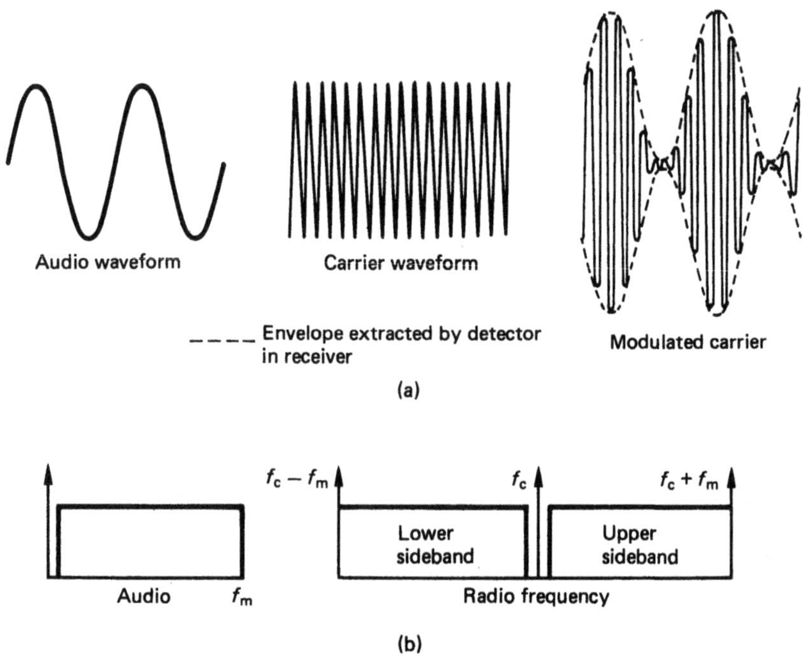

Figure 6.2 Amplitude modulation. (a) Amplitude modulation of RF carrier (100% modulation). (b) Spectrum of amplitude-modulated carrier

50 kW and above, an overriding consideration is to achieve a high efficiency in the conversion of input mains power to output RF power. The efficiency depends, mainly, on the type of circuit used to modulate the carrier and there are two basic systems, known as 'low-level' and 'high-level' modulation, which are illustrated in Figures 6.3(a) and (b). It is worth noting, however, that over the years the development of valves with increasing gain has contributed to a steady increase in efficiency. This is because fewer amplifier stages are required to achieve a given output power, with a consequent reduction in valve anode and heater power losses.

For a discussion of LF, MF and HF transmitter designs it is useful to summarize the types of amplifier circuits used. It should also be noted that, when the output of an amplifier is an amplified version of the input, the amplifier is described as 'linear'.

The *class A* circuit is a linear amplifier used at very low power levels in which the anode current (or transistor collector current) flows at all times. The anode efficiency is low because the anode loss is independent of the signal level.

The *class B* circuit is a linear amplifier in which the anode current flows for approximately half the cycle of the input signal. To remove harmonics an RF amplifier has a tuned load and an audio amplifier has two valves operated in push–pull. The anode efficiency is higher than for class A but does not exceed about 50%.

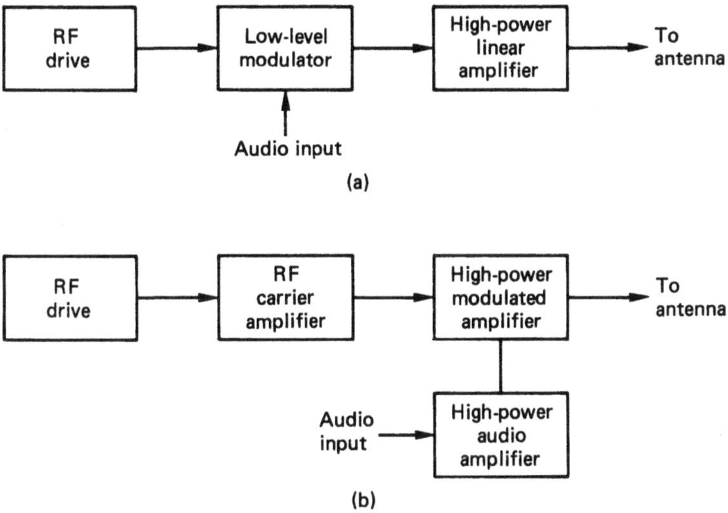

Figure 6.3 LF/MF and HF modulation systems. (a) Low-level modulation. (b) High-level modulation

The *class* C circuit is an RF amplifier in which the anode current flows for a small part of the cycle of the input signal. It cannot be used as a linear amplifier but the anode efficiency is high (over 80%).

6.2.1 Low-level modulation

As shown in Figure 6.3(a), in a low-level modulation system the complete modulated RF signal is generated at a low power level and then amplified up to the final output power in a class B linear amplifier. Since the early days of radio broadcasting this system attracted designers on account of its simplicity but in its basic form it has the disadvantage of low efficiency. This is because a class B linear amplifier can only operate efficiently at saturation (maximum anode current and minimum anode volts) and this occurs only when maximum signal is applied. As a result, the overall transmitter efficiency is in the region of only 30%, which means, of course, that for every 1 kW radiated, more than 2 kW of the input mains power is dissipated as heat. Here it should be noted that overall transmitter efficiency is conveniently expressed as the ratio of RF output power at average modulation depth (30%) and total input mains power.

As early as the 1930s, proposals were made for linear amplifiers with increased efficiency, and the most successful of these designs was due to W.H. Doherty. The Doherty linear amplifier is illustrated in Figure 6.4 and uses two valves, one carrying RF signals with an amplitude up to the level of steady carrier and the other carrying the peaks of modulation. High efficiency is achieved at 0% and

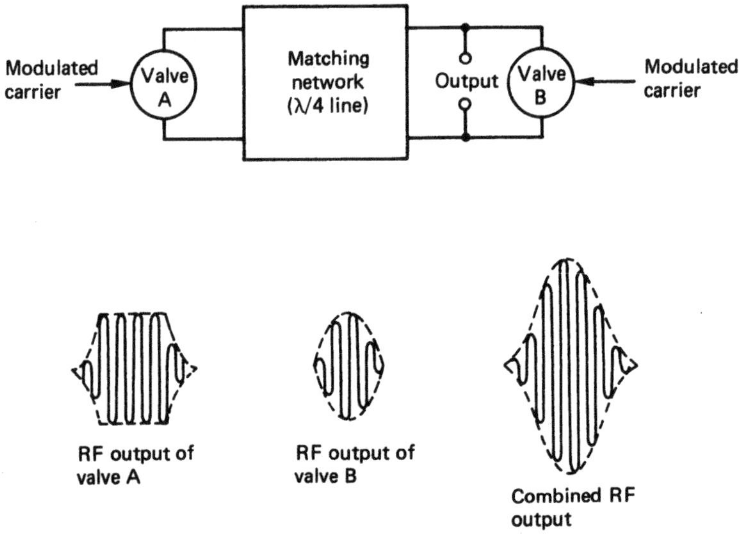

Figure 6.4 Doherty linear amplifier

100% modulated because, under both these conditions, the two valves operate near saturation. Under average modulation conditions the efficiency is lower and overall transmitter efficiencies between 50% and 60% can be achieved. As will be seen, the low-level modulation with a Doherty linear amplifier offers a significant improvement in efficiency over that with a class B linear amplifier and it is still used in transmitters with output powers up to about 50 kW.

6.2.2 High-level modulation

The Doherty linear amplifier requires careful lining up to avoid audio distortion and an alternative, also developed in the 1930s, is the high-level modulated transmitter illustrated in Figure 6.3(b). A commonly used form of high-level modulated transmitter is the class B anode-modulated arrangement which avoids the use of a linear amplifier and is suitable for very high power operation.

(a) Class B anode modulation

The class B anode-modulated arrangement is shown in Figure 6.5 and uses a high-efficiency class C RF output valve driven with the carrier frequency and operating under saturation conditions. The RF output of a class C RF amplifier varies linearly with the anode voltage so that if this is varied in accordance with the programme information, a high-power modulated signal is obtained. As shown in Figure 6.5, the anode voltage is varied by the insertion of the

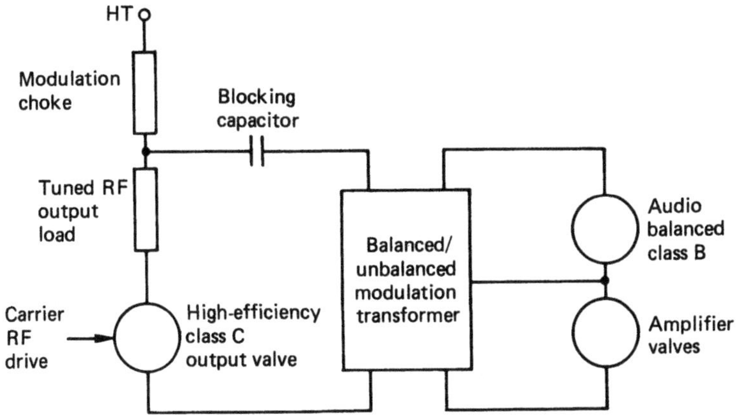

Figure 6.5 Class B anode-modulated system

modulating signal by a high-power balanced class B audio amplifier and output transformer.

As in the case of the Doherty linear amplifier, the efficiency of a class B anode-modulated transmitter is a maximum of 0% modulation and 100% modulation and drops significantly at average programme modulation levels due to losses in the audio amplifier. The overall efficiency is between 50% and 60% and is therefore similar to that of the Doherty arrangement. Within these limits, the actual overall efficiency tends to rise as the transmitter output power increases because the fixed losses (input power to the drive stages, etc.) become a smaller proportion of the total power consumption. Typical overall efficiencies of class B anode-modulated transmitters are shown in Table 6.1.

The disadvantage of the class B anode modulation system is that two large iron-cored components are needed. The modulation transformer that converts the balanced audio input to an unbalanced output must deliver a power equal to

Table 6.1

Carrier power (kW)	*Efficiency (%)*	
	0% Modulation	*30% Modulation**
50	62	56
100	64	58
250	65	59
500	66	60
1000	66	60

*Approximately equivalent to average programme modulation.

half that delivered by the HT (high tension) supply. The modulation choke that is necessary to remove the DC feed from the secondary of the transformer (thus avoiding magnetic saturation leading to non-linearity) is also large, as it must withstand high anode voltages and carry the DC feed corresponding to the steady carrier condition.

(b) Pulse-duration modulation (PDM)

The large iron-cored components required in the class B anode modulation system are expensive and occupy a considerable ground area. Further, the overall efficiency falls under programme modulation conditions. As a result, a compact arrangement and a very high efficiency cannot be achieved. During the 1970s an alternative high-level modulation system known as PDM (pulse-duration modulation) was developed and offers increased efficiency together with a more compact transmitter layout. The basic principles of PDM are shown in Figure 6.6 and it will be seen that the arrangement comprises a modulator valve in series with an audio frequency (AF) filter and a high-efficiency class C RF output valve. The AF programme signal is first converted into a constant-amplitude square wave with a pulse repetition frequency around 100 kHz. The pulses have a duration dependent on the amplitude of the AF signal and are used to bias the grid of the modulator valve so that it is either conducting with minimum losses or is open circuit with no losses. As a result, the modulator efficiency is high. The AF filter converts the square wave at the modulator anode to the original AF waveform and the diode maintains the current when the modulator valve is 'shut off'. Thus, unlike the other systems discussed, the efficiency of a PDM transmitter remains high at all modulation levels from 0% to 100% modulation.

Typical overall efficiencies of PDM transmitters are shown in Table 6.2.

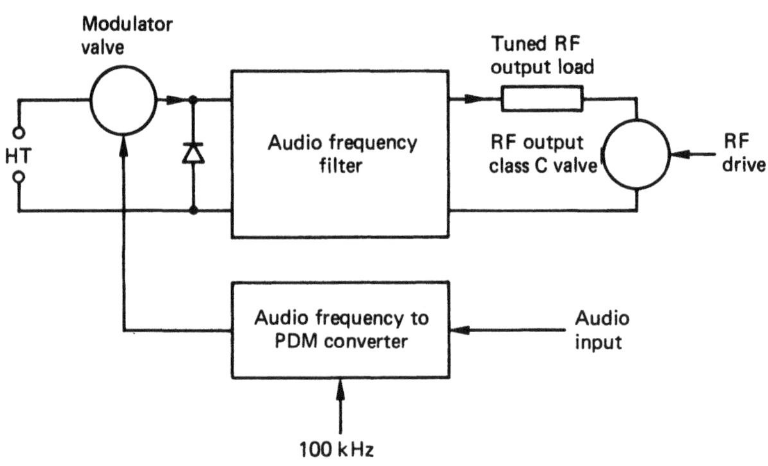

Figure 6.6 PDM anode-modulated system

Table 6.2

Carrier power (kW)	*Efficiency (%)*	
	0% Modulation	*30% Modulation*
50	62	63
100	63	64
250	64	65
500	65	66
1000	66	67

As will be seen, the efficiency at 30% average programme modulation is substantially higher than that of the class B anode modulation system shown in Table 6.1.

Summarizing, therefore, the three modulation circuits used in DSB AM transmitters have the following characteristics:

1. Low-level modulation with a Doherty linear amplifier results in a compact arrangement and a reasonably high efficiency. However, for optimum performance, careful lining up is necessary and the design is most suitable for fixed-frequency applications.
2. The class B anode modulation system is easy to line up and is well suited to applications where operational frequency changes are required. The disadvantage is that large and heavy iron-cored modulation components are required and militate against a compact layout.
3. The PDM (pulse-duration modulation) system has a higher efficiency than the other two systems and can be realized in a compact layout. PDM transmitters are suited to applications requiring operational frequency changes.

6.2.3 Dynamic carrier control (DCC)

In a conventional DSB transmitter the carrier amplitude is varied (i.e. modulated) by the programme signal. The steady carrier amplitude (i.e. the amplitude in the absence of programme modulation) must be such that, when fully modulated, it falls to zero and rises to twice the steady value at 100% modulation as shown in Figure 6.2(a). If the steady carrier amplitude is less than this value, distortion of the received programme signal will occur at high modulation levels as shown in Figure 6.7.

For modulation levels less than 100% the carrier level required to avoid distortion of the received signal is reduced. For example, at 30% modulation a carrier level of 30% of the full value will avoid distortion. It is theoretically possible, therefore, to reduce the input mains power by reducing the carrier

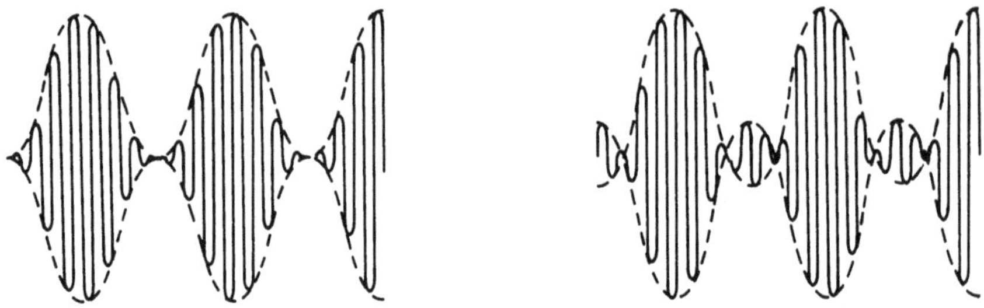

Figure 6.7 Over-modulation distortion

amplitude at levels of modulation below 100%. For example, with a conventional 100-kW transmitter the carrier level will be 100 kW at all modulation levels and, at an average modulation level of 30%, the total RF output power will be 105 kW. However, if at 30% modulation the carrier level is reduced to 30% of its full level the total power output will reduce to 14 kW and the mains input power will be correspondingly reduced. Unfortunately a reduction in input mains power of this order is impossible for a number of practical reasons but, during the 1970s and 1980s, circuits were developed that give worthwhile improvements in efficiency. These circuits go by a number of names (e.g. dynamic amplitude modulation and dynamic amplitude carrier modulation) but a convenient general designation is DCC.

The fundamental difficulty of DCC is that reduction of the carrier at lower levels of modulation is, in effect, a form of distortion. All receivers include AGC (automatic gain control) so that the effect of fading is minimized and the listener does not have to adjust his volume control when tuning between stations. Receiver AGC operates by changing the gain of the receiver as the average level of the received RF signal varies (e.g. as the signal level rises or falls, the gain is decreased or increased so the audio output signal remains at constant amplitude). Thus, if the carrier level is reduced with decreasing modulation, the gain will be increased so that the audio signal is compressed (i.e. the dynamic range reduced) and the background noise level increases. This represents distortion of the audio signal and limits the amount by which the carrier, and consequently the mains input power, may be reduced.

As early as the 1930s unsuccessful attempts were made to reduce the mains power consumption by carrier control but, in recent years, advances in microprocessor techniques have made DCC practicable. One approach is based on a reduction of carrier level with decreasing modulation depth and the RF resulting waveforms are illustrated in Figure 6.8(a). However, an alternative method developed by the BBC uses a conventional carrier level at zero modulation, and the complete transmitter output (carrier and sidebands) is then compressed at higher modulation levels so that the peak output power remains constant, as illustrated in Figure 6.8(b). The action of receiver AGC affects the AF output differently for these two methods of DCC. In the first method, as

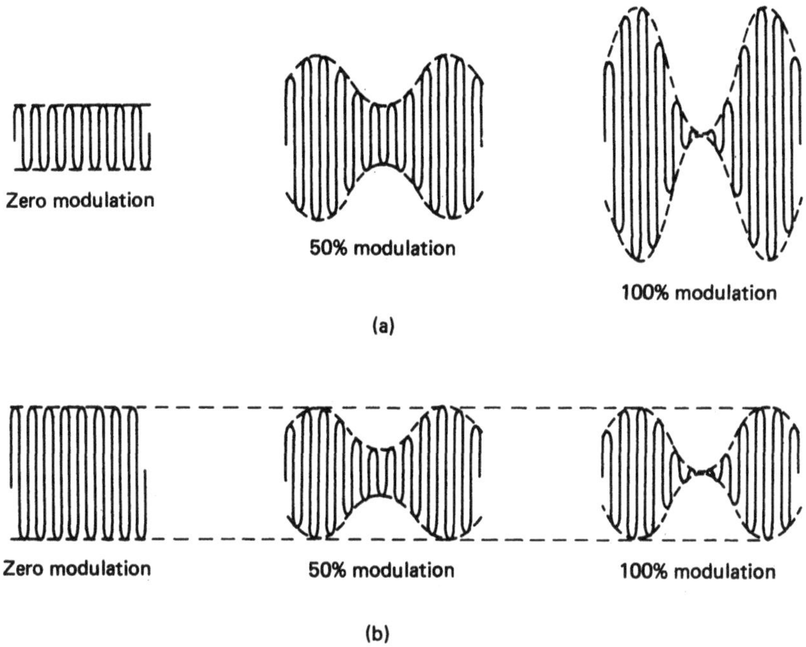

Figure 6.8 Alternative methods of DCC: RF waveforms. (a) Dynamic carrier control. (b) BBC compression method

described above, the receiver AGC results in a reduced dynamic range and an increase of audio background noise in quiet programme passages. However, in the BBC system, the mean signal level falls with increasing modulation depth. The receiver AGC therefore tends to correct the audio distortion resulting from modulation compression occurring at higher modulation levels. Further, since a full level carrier is transmitted at zero modulation depth, there is no increase of audio background noise in quiet programme passages.

DCC results in audio distortion which limits the saving of mains power that can be achieved, but with both the methods described above a saving in input mains power of between 15% and 20% is possible. For example, if a conventional 100-kW transmitter had an efficiency of 65% the input mains power of a comparable DCC system would have an efficiency of about 76%, which obviously represents a very worthwhile revenue saving.

6.3 LF and MF transmitter design

6.3.1 Solid-state transmitters

Since the 1960s there has been a steady increase in the power available from a solid-state output stage and all-solid-state transmitters with output powers in excess of 100 kW are commercially available. The difficulty in solid-state design is that the power available from a single transistor is limited and high output powers can only be achieved by the operation of a number of output modules in parallel with a resulting increase in circuit complexity. But there is little doubt that the power available from a single transistor will increase and the power output of all-solid-state transmitters will continue to rise. However, all-solid-state designs with very high powers in the megawatt (1 MW equals 1000 kW) range are becoming practicable, and are now being offered commercially.

The advantages offered by an all-solid-state transmitter are considerable. Such a transmitter has high reliability, since it can be designed so that in the event of a failure of one output module the remaining modules continue to operate and there is only a small reduction in output power. Thus there is no need to provide a standby transmitter. Further, the maintenance problem is simplified, as a faulty transistor module can be readily changed and, with appropriate design, a 'hot change' (i.e. without switching off the transmitter) is possible. Because there is no consumption of energy in valve heaters, the overall efficiency can be as high, and possibly higher, than that of a comparable transmitter with a valve output stage. The absence of heater power and the spread of waste heat in the output stage among a number of modules simplifies the cooling problem and a simple air-blast system can be used. Finally, an all-solid-state design has an important advantage as regards the risk of electrocution, as the supply voltage required is low (typically 30 V), but the risks of very heavy currents which low impedance sources can deliver, if accidentally short-circuited, should not be overlooked.

Efficiency is of less importance at lower transmitter power outputs but many all-solid-state designs use either the Doherty linear amplifier or a PDM system so that the modulation choke and transformer required for the class B anode modulation system are not required. A compact and economic design is easy to achieve: for example, a typical 10-kW all-solid-state transmitter has overall dimensions of 200 cm high by 180 cm wide by 65 cm deep together with an overall efficiency in excess of 80%.

6.3.2 Transmitters with tetrode output stages

Higher power transmitters use solid-state low- and medium-power stages so that valves (normally tetrodes) are only used in the output stage. An increasing number of designs use a PDM modulator and the problems encountered with this system at very high powers have been overcome in different ways by different manufacturers. The Marconi Pulsam system, the AEG Telefunken Pantel system and the Continental system use a switching valve in series with the output RF amplifier valve as in the basic arrangement of Figure 6.6. The Brown Boveri PSM system uses successive switching of solid-state power modules with increasing modulation depth so that a switching valve is not required. All these variations on the theme of PDM give high overall transmitter efficiencies. Also, transmitter designs using DCC in one form or another are available commercially and some manufacturers offer an add-on unit that can be used to improve the efficiency of some types of existing transmitter.

To increase coverage, there has been a steady increase in the output power of MF transmitters, and designs with output powers up to 1 MW and above are available. These high output powers have been made possible by the use of valves with pyrolytic graphite grids and improved methods of valve construction and cooling. Pyrolytic graphite grids can be precision-cut by sandblasting or laser-cutting, which results in precise, repeatable geometry and up to three times the power dissipation of a conventional grid of the same dimensions.

For an overall transmitter efficiency of (say) 70% the anode efficiency of the output valve can be as high as 90% but this still represents a significant amount of waste heat energy that must be removed by the valve cooling system. For example, in the case of a 500-kW transmitter the waste heat energy to be removed from the anode will be 30 kW or more.

Air-blast cooling is simple but is only practicable for transmitters with output powers up to about 100 kW and is often acoustically noisy. For higher powers, the heat from the anode of the output valve must be removed by circulating a liquid coolant (usually purified water) round the anode. For many years a liquid method was used in which the coolant was pumped round the anode and its temperature raised by removal of the waste heat. The coolant was then re-circulated after removal of the waste heat by means of a heat exchanger. The disadvantage of this technique is that a considerable flow of coolant is required to remove 50 kW or more of heat energy and this necessitates the provision of large-diameter pipework, pumps, heat exchangers and coolant storage facilities. Thus cooling of the output valve of a high-power transmitter by circulating a liquid coolant inevitably involves a bulky arrangement, which is expensive to run.

The size of the valve cooling system can be reduced dramatically by the use of the vapour-phase method (often referred to as 'steam cooling'), which is illustrated in Figure 6.9. The anode of the valve is surrounded by a boiler containing demineralized water and the level is kept constant by a level tank with a level-maintaining valve. The water at the anode is allowed to boil and the heat removed is absorbed by converting water into steam. The steam rises into a heat exchanger (a radiator cooled by an electric fan) where it is condensed and returned to the level tank and boiler and then re-circulated. The latent heat of vaporization of water is much greater than the product of specific heat and difference in inlet and outlet water temperatures in a liquid cooling system where the water is not allowed to boil. As a result, for a given anode dissipation, the flow of coolant in a steam cooling system is about one-fortieth of that required for a liquid cooling system. This results in a reduction in system pressures, the size of heat exchangers, pipework, pumps and the volume of coolant to be stored.

The effectiveness of a vapour-phase cooling system depends critically on the design of the valve anode. If heat is dissipated in an anode with a smooth metal

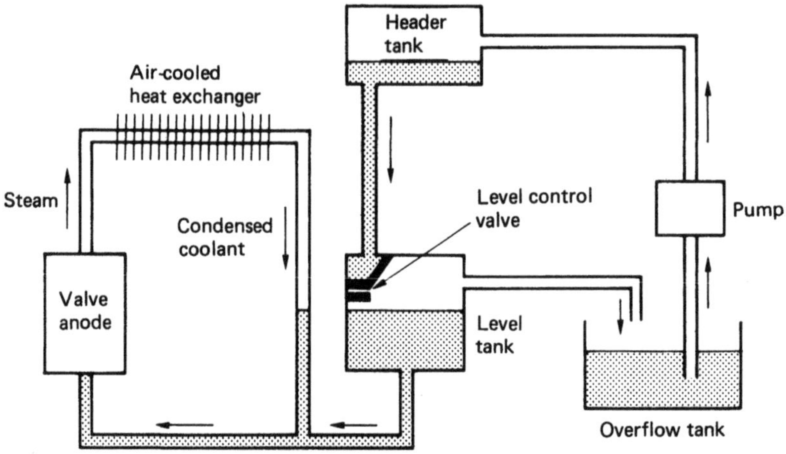

Figure 6.9 Vapour-phase valve cooling system

surface at atmospheric pressure, the water commences to boil at 100°C. At surface temperatures above 100°C the heat transferred to the water rises rapidly, reaching a maximum of about 150 W/cm^2 at a temperature of 125°C. When the surface temperature rises above 125°C an insulating film of steam forms around the surface (a phenomenon known as 'cavitation'). The heat cannot then escape from the anode, so its temperature rises rapidly and it may be destroyed. To dissipate more than 150 W/cm^2 it is necessary to inhibit the formation of an insulating film of steam and this can be achieved by using an anode with surface protrusions. The protrusions cause turbulence in the water which prevents the formulation of a film of steam, and heat transfer from the anode can be in excess of 500 W/cm^2.

6.3.3 Maximum transmitter power

To increase the range covered by a transmitting station it is, of course, necessary to increase the ERP (effective radiated power, the product of transmitter power and antenna gain). In the MF band, unlike the VHF band, radiating elements cannot be stacked vertically and the gain of both omnidirectional and directional antennas is therefore limited to a few dB. Since high antenna gains cannot be achieved, the only effective means of increasing the ERP is to increase the transmitter power. As a result there has been a steady increase in MF transmitter power over the past decades and powers up to 2 MW are not uncommon. However, very high power MF transmitters are very expensive from the capital and revenue points of view and it is necessary to investigate the 'trade-off' between increasing transmitter power and the resulting increase in service area.

Table 6.3

Radiated power (kW)	*Range to field-strength contour (km)*					
	600 kHz		1000 kHz		1600 kHz	
	85 dBμ	66 dBμ	85 dBμ	66 dBμ	85 dBμ	66 dBμ
10	29	91	18	54	12	35
50	45	128	27	76	17	47
100	53	147	34	88	20	57
250	70	175	40	105	24	66
500	80	200	45	120	29	76
1000	95	225	55	140	33	88

In MF domestic transmission the ground-wave is used and the field strength decreases with increasing frequency and with decreasing ground conductivity; both these parameters have a profound effect on propagation. For example, assuming the same transmitter power, the range of coverage in kilometres with a low frequency (600 kHz) and a high conductivity (10 mS/m) is approximately ten times that for a high frequency (1600 kHz) and a low conductivity (1 mS/m). In considering the effect of transmitter power on coverage it is therefore necessary to take frequency and conductivity into account. However, to avoid undue complication, an average ground conductivity of 5 mS/m will be assumed. Table 6.3 summarizes, approximately, the range of coverage obtained with radiated powers ranging from 10 kW to 1000 kW (1 MW). The field strength of 85 dBμ will give coverage of a large town or city and the field strength of 66 dBμ will give acceptable rural coverage.

Table 6.3 shows that, on average, the coverage range increases by only about 15% if the transmitter power is doubled. At low transmitter powers the cost of such an increase can be acceptable but at high powers, where a doubling of power involves a heavy financial penalty, the increase in range is much less likely to be cost-effective. Further, Table 6.3 shows that, at high frequencies, the increase in range resulting from a large increase in transmitter power only amounts to a few kilometres.

Table 6.3 ignores the effect of the sky-wave which is radiated by an MF antenna at angles above the horizontal and is reflected back to earth at night. As explained in Chapter 2, the sky-wave varies in both magnitude and phase and, at distances from the transmitter where its amplitude is comparable to that of the ground-wave, fading and distortion of the received signal occur in an annular area known as the 'fading zone'. For a quarter-wave antenna the radiated field is high at high elevation angles and the inner limit of the fading zone is relatively near the transmitter. However, if an antenna with a height around a half-wavelength is used (an 'anti-fading' antenna) the radiation at high elevation

Table 6.4

Frequency (kHz)	*Distance of inner limit of fading zone (km)*	
	Quarter-wave antenna	*0.55 Wavelength antenna**
600	98	156
1000	108	162
1600	78	117

*0.55 wavelengths is the optimum height for an anti-fading antenna.

angles is reduced and the inner limit of the fading zone is independent of transmitter power; for the frequencies shown in Table 6.3, the distances from transmitter are shown in Table 6.4.

As has been noted in Chapter 2, the MF sky-wave can provide extensive coverage during the hours of darkness. Unlike ground-wave propagation, the sky-wave field strength is substantially independent of frequency and does not decrease, steadily, with increasing distance from the transmitter. In fact, the sky-wave is zero in the vicinity of the transmitter (due to the zero vertical radiation from a vertical antenna) and then increases to a maximum with increasing distance before decreasing as the distance is further increased. As a result there is a minimum transmitted power that will provide a field strength in excess of the 73 dBμ normally assumed necessary for sky-wave reception. Sky-wave broadcasting is usually aimed in a particular direction, so it is normal to use a directional antenna. Table 6.5 shows the sky-wave coverage, as a function of frequency, for the typical case of a two-mast directional antenna and the mid-band MF frequency of 1000 kHz.

Table 6.5

Transmitter power (Kw)	*Sky-wave coverage (field-strength greater than 73 dBμ)*	
	from	to
10	*	*
50	*	*
100	230 km	320 km
250	230 km	580 km
500	230 km	780 km
1000	200 km	1000 km

*Field strength less than 73 dBμ at all distances.

It will be seen that there is no useful sky-wave coverage for powers below 100 kW and that to achieve a service at a distance of 1000 km a transmitter power of at least 1 MW (1000 kW) is necessary.

There are two approaches to the domestic MF broadcast coverage of large geographical area. A network of medium- and low-power transmitting stations can be used or, alternatively, a few high-power stations. However, Table 6.3 makes it clear that even very high powers give relatively limited ground-wave coverage at high frequencies. For example, at 1600 kHz, even 1 MW will only give city coverage (85 dBμ and above) out to a distance of 33 km and rural coverage (greater than 66 dBμ) out to 88 km. It follows that a high transmitter power is only likely to be cost-effective when a frequency low in the MF band is available.

(a) Coverage of densely populated industrialized countries

In this case, a large part of the area of the country will require a field strength giving 'city coverage' (85 dBμ). However, even with a low MF frequency, Table 6.3 shows that a transmitter power of 1 MW will only give this field strength out to a range of about 90 km. High-power transmitters, therefore, will not be cost-effective and a network of medium- and low-power stations (with a power ranging from around 5 to 100 kW), and sited at or near centres of population, will provide the most economic coverage.

(b) Coverage of thinly populated Third World countries

Here the level of electrical interference is generally low and a rural field strength (66 dBμ) will usually give acceptable reception even in small towns. Further, a lack of infrastructure facilities (power supplies, road access and programme feed) will probably make an extensive network of medium- and low-power stations difficult to achieve. Thus there is a case for high MF transmitter powers which, with a low MF frequency, will give the maximum possible area coverage. However, Table 6.4 shows that, even using an expensive anti-fading antenna, the fading-free night-time coverage is, at most, around 160 km and, from Table 6.3, this can be achieved with a transmitter power of about 200 kW. Whether a higher power will be cost-effective will depend on the size of the country involved and the population spread within that country. It may be that the fading zone occurs in areas of very low population and this may justify the use of transmitters with powers above 200 kW (i.e. up to 2 MW).

As can be seen in Table 6.5, there is a significant increase in long-distance sky-wave coverage with increasing transmitter power. Thus the power required for a sky-wave transmission will depend on the distance at which it is required to provide a service. As has already been pointed out, a minimum power of 100 kW is required, a transmitter power of 1 MW will provide a service out to a distance of 1000 km, and higher powers will provide greater ranges.

Summarizing, the conclusions that can be drawn regarding the relationship between MF transmitter power and service area are:

1. For domestic broadcasting in highly populated industrialized countries the most effective arrangement is a network of stations with transmitter powers up to about 100 kW.
2. For domestic broadcasting in thinly populated Third World countries a lack of infrastructure facilities may make a few high-power transmitting stations more attractive than a network of lower power stations. However, it is preferable to use frequencies low in the MF band and, even then, fading may limit the maximum useful transmitter power to around 200 kW.
3. For long-distance sky-wave coverage very high transmitter powers can be used and the optimum transmitter power will depend on the distance at which a service is to be provided.

6.4 HF transmitter design

6.4.1 Design considerations

The principal difference between DSB HF and DSB LF/MF transmitters is that the former must be capable of the rapid frequency changes that are needed to maintain a service over the ionospheric diurnal, seasonal and sunspot cycles. Apart from this requirement, developments in the HF transmitter design have followed those in the LF/MF field as regards output power, increased efficiency and smaller size. However, an exception is that the Doherty linear amplifier is not suitable for HF operation as it is difficult to tune for changes in operating frequency in the short time available (typically less than 12 seconds). The maximum transmitter output power normally available is 500 kW and tetrode output stages with PDM modulators are increasingly common. The 500-kW maximum transmitter power is dictated partly by available valve designs but mainly by the difficulty of designing high-gain antennas that will carry the very high peak voltages associated with higher powers. Early designs of HF transmitter required manual changing of the tank-circuit components but all current designs include automatic, remote-controlled frequency changing. Control circuits commonly use a combination of solid-state and relay logic, and start–stop control commands are provided for all transmitter functions. Transmitter and antenna control and monitoring allow remote or extended control from a video terminal. Multi-channel operation is provided with channel frequencies stored in the control computer memory. Also, thousands of operating logs can be stored and reviewed. For several decades, the trend in transmitting station design has been increasingly to unattended operation although the high voltages involved and the complexity of the frequency and antenna switching make it difficult to envisage a complete withdrawal of all maintenance staff at high-power multi-power HF transmitting stations. However, it is worth remembering that there is still a role for simple HF transmitters with semi-manual operation for tropical broadcasting applications and in cases where highly trained maintenance staff are not available.

6.4.2 Transmitter output power

It has already been pointed out that in the LF and MF bands very high output powers are only cost-effective in a small number of applications. But the HF case is different since the coverage of an HF transmitter is not limited by factors unrelated to transmitter power (high ground attenuation and the creation of a fading zone). As a result, any increase in transmitter power will result in a useful increase in field strength and, theoretically at any rate, there is no obvious limit to the maximum power that could be used. However, there is a limit to the maximum useful transmitter power in the HF band which is set by practical considerations concerning antenna power-carrying capacity. This limitation is fundamental and can only be overcome by the use of larger diameter conductors which would pose severe structural design problems. The highest HF transmitter output power normally available is 500 kW, and if there is any increase it is unlikely to be more than up to 1 MW. Even this maximum power appears unlikely as, on the occasions that a very high power is required, it would be more economic to drive a 'split' curtain antenna with two 500-kW transmitters as illustrated in Figure 6.10.

6.4.3 The introduction of SSB HF broadcasting

As shown in Figure 6.2(b), the output of a DSB transmitter comprises the steady carrier term and sidebands both above and below the carrier frequency (the

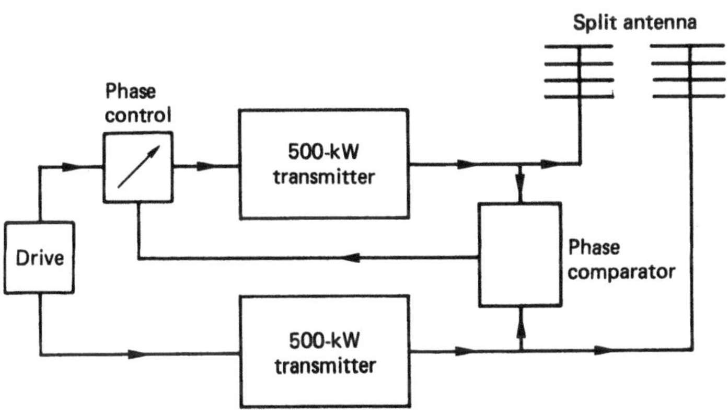

Figure 6.10 Two-transmitter/split antenna arrangement

upper and lower sidebands). The upper and lower sidebands each contain all the programme information, which is therefore duplicated, and the steady carrier term obviously carries no programme information. It follows, therefore, that one set of sidebands together with the steady carrier term can be regarded as redundant and the transmitter output power they represent can be regarded as wasted. It is, therefore, at first sight only necessary to transmit a single set of sidebands (SSB transmission) with a consequent reduction of the transmitter mains input power required. However, it is necessary to transmit a low-level carrier (requiring little input mains power) to provide the receiver with a carrier reference, as a carrier voltage in the correct phase is required in the receiver to enable demodulation to take place.

SSB broadcast transmission has important advantages over DSB transmission:

1. Taking a 100-kW DSB transmitter as an example, the carrier power is 100 kW and the maximum power in each set of sidebands (i.e. at 100% modulation) is 25 kW peak effective power (PEP). If only one set of sidebands is transmitted, the demodulated audio signal will be reduced by 6 dB but the RF bandwidth will be halved so that the received signal-to-noise ratio (SNR) will be reduced by 3 dB. It follows that an SSB transmitter with 50 kW PEP will provide the same service as a DSB transmitter with 100 kW carrier power. Under modulation conditions, the average RF output power of a 50-kW PEP SSB transmitter will be about 12 kW and that of a 100-kW DSB transmitter will be about 112 kW so that there is a reduction in the mains power required of up to 10:1.
2. SSB transmitters can be designed to transmit two different programmes, one occupying the upper and one the lower set of sidebands, and this is termed ISB (independent sideband broadcasting). The number of broadcast channels in the HF broadcast band can be doubled if SSB operation is used.
3. In HF ionospheric transmission, the signal is often transmitted to the receiver by more than one path and this can result in cancellation of the carrier and/or a sideband frequency. This is known as selective fading and can cause severe distortion which (since the carrier amplitude is unimportant and there is only one set of sidebands) can be reduced, considerably, with SSB transmission.

A SSB receiver is more complex than a DSB receiver, as in the former the carrier term must be regenerated before demodulation can take place. The low-level carrier term that forms part of the SSB transmitted signal can be used to control an oscillator in the receiver and thus regenerate the carrier at high level and in the correct phase relationship with the transmitted sideband. The general arrangement of a SSB receiver is shown in schematic form in Figure 6.11 and it should be noted that cheap and compact designs have been made possible by the introduction of solid-state techniques.

In view of the advantages of SSB, particularly as regards the eventual doubling of the number of broadcast channels, the World Administrative Radio Conference (WARC) has resolved that it should be introduced into the HF broadcasting bands. To allow a phasing-out period for DSB receivers, the

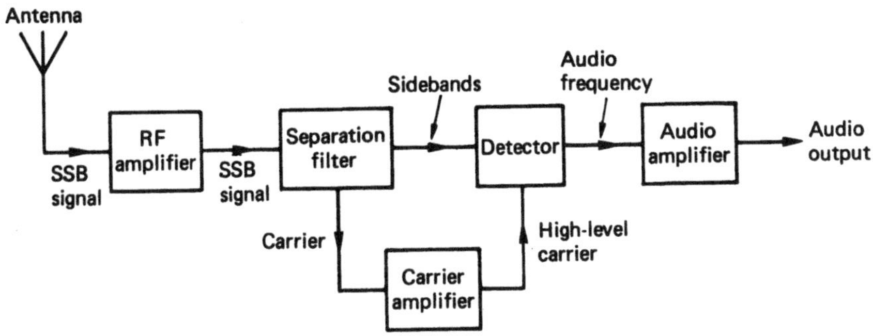

Figure 6.11 SSB receiver

transmitted carrier level will be reduced by 6 dB as soon as is practicable. This will cause some distortion to reception with DSB receivers but, in view of the normal quality of HF transmission, the degradation is expected to be acceptable. Then, and not later than 31 December 2015, the transmitted carrier will be reduced by 12 dB (to the agreed level for SSB broadcasting). To meet this programme, all new transmitters should be capable of transmitting SSB with the alternatives of a 6-dB or 12-dB reduction in carrier level.

It is worth noting that, some years ago, a compatible single sideband system (CSSB) was proposed that would, in theory, enable SSB transmissions to be made that could be received using a conventional DSB receiver. Such a system, however, has the disadvantages that a full carrier would need to be radiated (thus losing the advantage of reduced transmitter mains power consumption) and an increased bandwidth would be required (thus reducing the number of additional channels created by the introduction of SSB). As a result, it is unlikely that broadcasting authorities will wish to use the CSSB system.

SSB transmitters for communications links have been available for many years but the application of SSB to broadcasting gives rise to a number of problems:

1. Traditionally, SSB and ISB communications HF transmitters have used a low-level modulator followed by a linear power amplifier. Such an arrangement has a low overall efficiency but in view of the lower mains power input required for SSB, and the relatively low radiated powers used for HF links, this is quite acceptable. However, SSB broadcasting to listeners with cheap receivers and simple antennas requires higher radiated powers and transmitter efficiency becomes important.
2. The WARC resolution to introduce SSB requires that the carrier level is reduced in two stages (first by 6 dB and then by 12 dB). Thus, ideally, transmitters should be capable of DSB operation and also of SSB operation with two different carrier levels.
3. There are many existing DSB transmitters (some brought into service relatively recently). Thus, to minimize the cost of the change to SSB it is desirable that as many as possible of these transmitters should be converted to SSB operation.

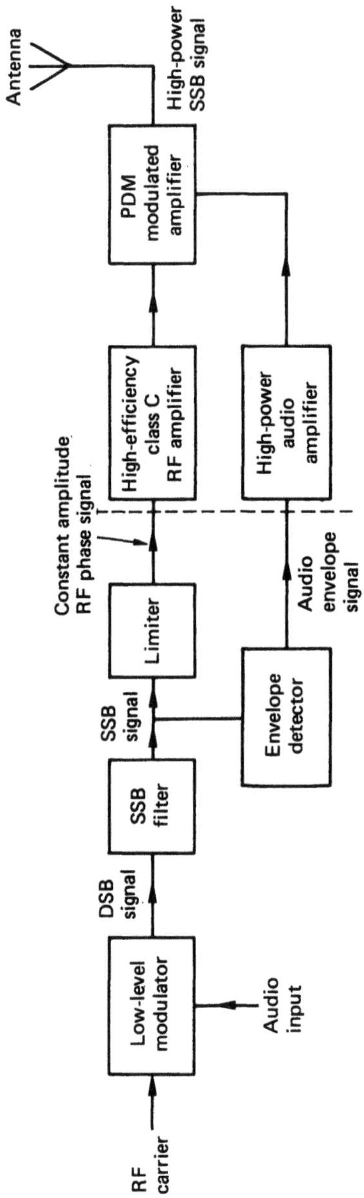

Figure 6.12 High-efficiency EER transmitter

A solution to the transmitter problems raised by these requirements is the EER (envelope elimination and restoration) technique proposed by Kahn and illustrated in schematic form in Figure 6.12. As will be seen, a low-power SSB RF waveform is first generated and the phase and audio amplitude modulation components are then separated. The phase term is obtained by limiting the RF waveform so that a constant-amplitude RF waveform is obtained that can be amplified in high-efficiency class C stages. The audio amplitude term is obtained by deriving the 'envelope' of the RF waveform and amplifying in a linear AF amplifier. This term is used to high-level modulate the amplified phase term using a PDM or class B anode modulator. It can be seen that, to the right of the dotted line in Figure 6.12, the transmitter arrangement is similar to that of a conventional high-level, high-efficiency transmitter. It is therefore possible to use different low-power stages (shown on the left of the dotted line in Figure 6.12) to obtain a range of transmitted signals that differ as regards carrier level and sideband distribution. Using the EER technique it is possible to design a high-efficiency transmitter that, by switching, can radiate DSB or SSB with different carrier levels. Further, by replacing the low-power stages it is possible to modify suitable existing transmitters to give the same range of transmitted RF signals.

An SSB transmitter with a PEP of 50 kW will give the same coverage as a DSB transmitter with 100 kW PEP carrier power. In other words, an SSB transmitter needs to have a PEP of only half that of the carrier power of a corresponding DSB transmitter. Further, the PEP of the DSB transmitter (with 100% modulation) is four times the carrier power so that an SSB transmitter has a PEP of only one-eighth of that of a DSB transmitter. It follows that an antenna suitable for operation with (say) a 100-kW DSB transmitter would be suitable for operation with a 400-kW PEP SSB transmitter, which would result in an increase of 9 dB (eight times) in field strength. Thus SSB offers the alternatives of a large reduction in transmitter mains input power or the possibility of a large increase in the field strength of HF broadcast transmissions.

Author's note:
Solid state modulation systems: It is worthwhile noting that during 1991–2 solid state modulation systems for high power MF and HF broadcasting transmitters became available. These modulators have extremely high efficiences (above 95%) as there are no large iron cored components or vacuum tubes in these designs. The solid state modulator consists of a number (normally 48) of series connected modules which are switched on or off to provide the high voltage DC and superimposed high level audio voltage to the main input tube.

Chapter 7

VHF and UHF radio and television transmission

7.1 Introduction

By the late 1940s the LF and MF broadcast bands were heavily used and, as a result, many services were limited by night-time co-channel interference. It was therefore necessary to bring a higher frequency band into use so that further transmissions could be accommodated. The VHF Band II (approximately 88–108 MHz) using frequency modulation (FM) was allocated to radio broadcasting and, besides providing additional channels, had a number of important advantages over the LF and MF radio broadcast bands. Because of the higher frequency band, much wider channels are available and this enables FM to be used so that programmes of higher quality, including stereophony (stereo), can be transmitted. Further, because VHF transmissions are not reflected by the ionosphere, co- and adjacent-channel interference are much reduced. Because of these advantages, it was at one time thought that MF transmissions could be phased out and replaced by transmissions in the VHF band. However, the invention of the transistor heralded the proliferation of portable and car radios which were more suited to the existing LF/MF services than the then favoured horizontally polarized (HP) VHF/FM transmissions, and the demand for radio broadcast services became so great that LF, MF and VHF transmissions are now required to provide sufficient channels. Other reasons for the retention of LF/MF broadcasting are that, when compared with VHF/FM, it is cheaper and (particularly at the lower MF and LF frequencies) can provide a greater coverage from fewer sites. However, in contrast with reasons for the retention of LF/MF broadcasting, there are incentives to introduce even higher quality services than can be provided by VHF/FM broadcasting. The greatly improved quality of VHF/FM reception by portable and mobile receivers resulting from the transmission of a vertically polarized (VP) component was noted in Section 4.2.2, but it is a fact that reception by such receivers, particularly in city and urban environments, is generally inferior to reception by fixed receivers employing directional receiving antenna systems which are able to minimize the effect of multi-path. With the introduction of CD – compact discs – in the early 1980s these deficiencies became more apparent and efforts were made to devise a system which would be equally good for fixed, portable and mobile reception. By the early 1990s a consortium of European partners had demonstrated such a system employing digital techniques, and it seems likely that services employing such techniques, referred to as DAB (digital audio broadcasting) will begin to be introduced in the mid-1990s. Receivers for these services can, however, be expected to be more expensive than those for LF/MF and VHF/UHF services, and will probably operate in the VHF band, though the system is also suitable for satellite broadcasting, or a mixture of the two.

In the 1930s high-definition television services commenced and this meant that frequency bands above those used for MF and HF radio broadcasting had to be introduced. Television transmissions have much wider modulation bandwidths than radio transmissions (several MHz as opposed to between 5 and 15 kHz) and can only be accommodated in the VHF and UHF bands. Television transmissions commenced in the VHF Band I (approximately 45–80 MHz) since, at these frequencies, transmitter and receiver designs could be based on existing HF techniques. As television services expanded, there was a need for additional channels and the VHF Band III (approximately 175–220 MHz) and the UHF Bands IV and V (approximately 470–900 MHz) were brought into use.

It will be seen, therefore, that the advent of VHF/FM radio and of high-definition television transmissions has made it necessary to bring new and higher frequency bands into use and this has had a profound influence on transmitter technology. To generate high powers in the VHF and UHF bands is much more difficult than in the LF and MF bands and it was necessary to develop new forms of thermionic tubes and higher power transistors. This development work has resulted not only in higher output powers but also in increased transmitter efficiency which, in view of the high cost, worldwide, of mains power is of great importance.

On a practical note, sometimes to meet the output power requirements or to ensure reliability of the service two transmitters are used. Worldwide there have been a number of fires in transmitter halls which have destroyed both transmitters. Hence, to the extent to which it is practicable, each transmitter and its ancillary items should be physically separated, in the limit, in separate rooms with fire barriers between.

7.2 VHF and UHF transmission systems

7.2.1 Radio broadcasting systems

(a) VHF/FM monophonic (mono) transmissions

In the FM system, the carrier frequency is varied in accordance with the amplitude of the audio programme signal and the carrier amplitude is held constant. This contrasts with the AM (amplitude modulation) system used in the LF and MF bands (described in Chapter 6) where the carrier frequency is held constant and its amplitude is varied. The spectrum of an FM transmission comprises a series of sidebands, above and below the carrier frequency, spaced by the modulating frequency, as illustrated in Figure 7.1. Although, theoreti-

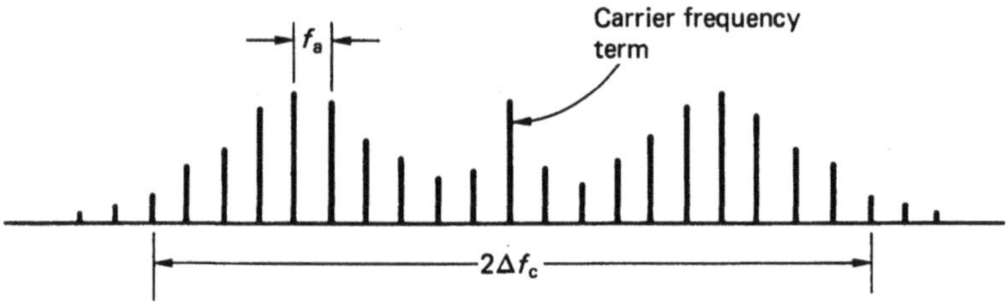

Figure 7.1 Spectrum of frequency-modulated transmission. f_a = audio modulating frequency; Δf_c = maximum frequency deviation of carrier

cally, there is an infinite number of sidebands, they fall rapidly in amplitude away from the carrier. As a result, distortion is avoided if the RF bandwidth is greater than twice the maximum frequency deviation or twice the maximum modulating frequency, whichever is the greater. By the maximum frequency deviation is meant the maximum amount by which the instantaneous carrier frequency is varied from its steady value (this is equivalent to 100% modulation in the AM case). The maximum modulating frequency for monophonic (mono) VHF/FM transmissions is the maximum audio frequency (AF) (15 kHz) and the maximum frequency deviation is ±75 kHz. Thus the RF bandwidth needs to be at least ±75 kHz and in practice is usually made equal to ±105 kHz to the '6 dB points' (see Appendix 2). This enables a spacing of 200 kHz between adjacent VHF/FM channels to be used without serious adjacent-channel interference.

A fundamental advantage of VHF/FM is that noise at the receiver output, when compared with that of an AM transmission of the same power, is lower by about 20 dB. This is because the noise affecting the received signal is predominantly amplitude modulated and, since the programme information is frequency modulated, this noise can be removed by limiting in the receiver as illustrated in Figure 7.2. It should also be noted that some noise reduction is obtained by increasing the level of the audio signal at the higher frequencies (pre-emphasis) with a corresponding reduction (de-emphasis) in the receiver.

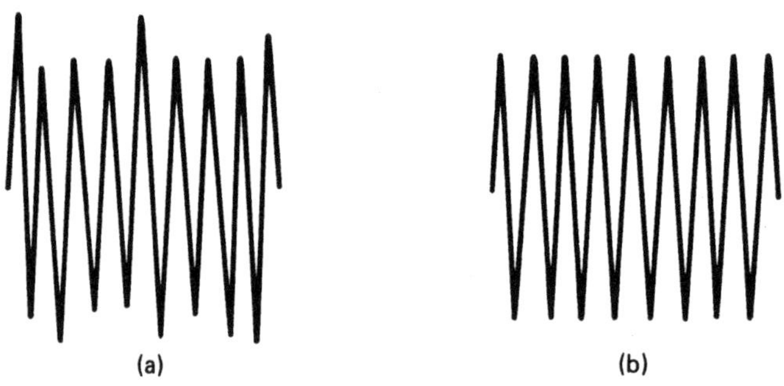

Figure 7.2 Removal of noise by limiting. (a) Carrier amplitude modulated with noise. (b) Carrier after limiting

The time-constant of the network used to provide the pre- and de-emphasis varies and may be 50, 75 or 150 μs (see Section A2.12).

(b) VHF/FM stereophonic (stereo) transmissions

From the relationship between frequency deviation, audio bandwidth and RF bandwidth given above, a $\pm$105 kHz RF bandwidth will accommodate a modulation bandwidth up to about 100 kHz. It will be seen, therefore, that, provided the maximum frequency deviation is kept below $\pm$100 kHz, the modulating frequency can be increased to a value approaching 100 kHz. Thus, without any increase in RF bandwidth, it is possible to transmit signals in the range above 15 kHz (the maximum audio programme frequency) and this makes stereo transmission possible in the same bandwidth as that required for mono transmission. It is important to note, however, that the system used for stereophony must be 'compatible', by which is meant that stereophonic transmissions can be received by a listener with a monophonic receiver without loss of quality. The VHF/FM stereo system now used worldwide is illustrated in Figure 7.3(a), which shows the frequency spectrum of the modulating signal (usually termed the 'baseband' signal). The programme material (from a studio or tape machine, etc.) comprises two audio signals defined as L (left) and R (right) and these are added and subtracted to provide two new signals, $(L+R)/2$ and $(L-R)/2$. The $(L-R)/2$ signal is then amplitude modulated on a 38-kHz sub-carrier and the carrier term is suppressed. It is a feature of the system that the complete baseband signal has a maximum amplitude equal to L or R, whichever is the greater, so that the RF frequency deviation is no greater than that of a mono system. In addition to the $(L+R)$ and modulated $(L-R)$ signals, a 19-kHz low-level carrier term is transmitted which is doubled in the stereo receiver to regenerate the 38-kHz carrier. In the receiver, the $(L-R)/2$ term can then be recovered by demodulation (using the regenerated 38-kHz carrier) and added to the $(L+R)/2$ signal to produce the L signal and subtracted to provide the R signal. A mono receiver uses only the $(L+R)$ term which, in practice, provides a service similar to that from a mono programme source.

A disadvantage of stereo reception is that the wider baseband introduces additional AF noise and, as a result, the received signal-to-noise ratio (SNR) is about 20 dB lower than that for mono reception with the same transmitted power. Except at receiving locations near the transmitter it is therefore necessary to use an external antenna at roof-top level to take full advantage of stereo transmissions, and so good stereo reception in cars is not usually feasible.

(c) SCA and control transmissions

International standards limit the maximum baseband modulating frequency to 99 kHz so that adjacent-channel interference is within acceptable limits. As a result, it is possible to transmit one or more additional programme channels in

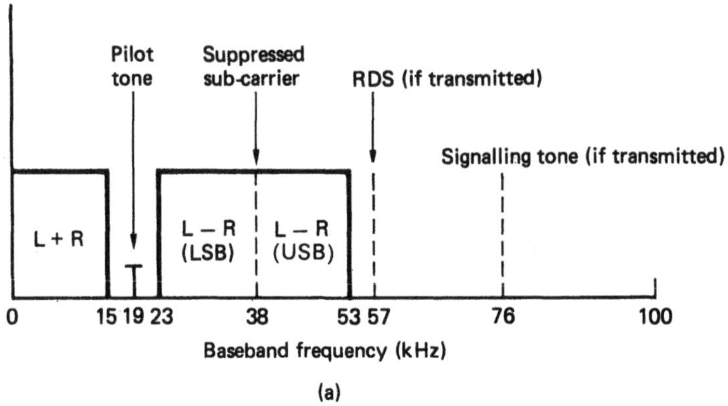

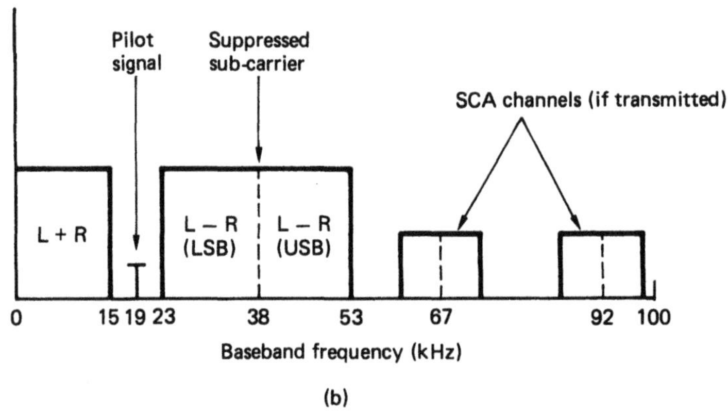

Figure 7.3 VHF/FM baseband spectra. (a) Stereo transmission baseband spectrum. (b) Stereo plus two SCA transmission baseband spectrum. L + R, sum of left- and right-hand audio signals (monophonic signal); L − R, difference of left- and right-hand audio signals (difference signal); LSB, lower sideband of difference signal; USB, upper sideband of difference signal

the baseband frequency range above 53 kHz (the maximum modulating frequency required for stereo transmissions). Such channels are usually referred to as SCA (subsidiary communications authorization) after their original American designation. SCA channels are carried on frequency modulated sub-carriers and, since they have restricted bandwidth and restricted frequency deviation, they are relatively low grade as regards bandwidth and SNR. SCA channels can, however, provide a programme quality comparable to that of an MF transmission and offer a means of providing additional mono programme channels or control or monitoring information. Transmitters employing FM re-broadcast reception (RBR) programme feeds (see Section 5.4.5) can be 'captured' if the normally received signal is swamped by a stronger signal: control signals can be used to prevent this, by shutting down the transmitter if an attempt is made to 'capture' it. Figure 7.3(b) shows the baseband spectrum of a VHF/FM channel, including stereo transmission with additional SCA channels.

(d) RDS transmissions

RDS – radio data system – developed by the BBC is a system which adds information in digital form to FM radio transmissions. The information is conveyed by phase modulating a 57-kHz sub-carrier (three times the frequency of the pilot tone) and so does not interfere with the programme signals. With a suitable receiver it offers fully automatic tuning to the strongest signal of a national radio programme of the listener's choice, which is a particularly desirable feature for mobile reception when driving out of the service area of one transmitter and into that of another. It can also provide other services, including giving an accurate display of time, interrupting programmes to give local traffic reports, displaying a limited amount of text and acting as a transparent data channel.

(e) DAB

Digital audio broadcasting (DAB) using coded orthogonal frequency division multiplex (COFDM) is a system of radio broadcasting which uses several hundred carriers, packed closely together in frequency, and permits several programmes to be transmitted simultaneously with the digital information for all the programmes spread over all the carriers as opposed to conventional systems when one carrier is associated with each programme.

The reception of digital transmissions can be seriously impaired if, in addition to the main signal, delayed signals, e.g. due to reflections from obstacles (multi-path), are present at the receiving location. The degree of impairment depends not only on the amplitude but also on the delay of any delayed signal relative to the bit-rate (pulse length) – the higher the bit-rate, the shorter the permissible delay. However, RDS transmissions are feasible because of the relatively low bit-rate, and the Teletext (described in Section 7.2.2) and NICAM transmissions (described in Section 5.6) associated with television because the directivities of receiving antennas minimize the effect of delayed signals.

In the DAB COFDM system digital information derived from the programmes is processed, by taking advantage of the characteristics of human hearing, to reduce the bit-rate to only slightly more than a quarter of that used for the earlier NICAM coding system and is then used to modulate the carriers using DPSK – differential phase shift keying.

The use of many carriers and the bit-rate reduction means that the bit-rate for each carrier is relatively low, which in turn means that delayed signals with relatively long delays can be tolerated, thus dispensing with the need for narrow-beam receiving antennas and permitting the use of quasi-omnidirectional antennas necessary for reception in vehicles. Indeed, the bit-rate is so low that it is feasible to use many transmitters operating on the same frequency to give national coverage (though it may be necessary to ensure that the transmitters are frequency locked and the digital signals delayed appropriately at each or some of the transmitting sites).

The reduced bit-rate results in lower bandwidth requirements and one proposed system requires only 4 MHz for national coverage of the UK with 12 stereo programmes, compared with a bandwidth requirement of about 2.8 MHz per programme with conventional FM.

Another advantage of the multi-carrier approach and coding system used is that since digital information (coded sound bits) for each radio service is spread across a number of carriers, interference to a particular carrier does not wipe out any of the services. Although all the digital information for all the programmes is spread over all the carriers, the method of coding and modulation enables the information for any programme to be faithfully recovered by suitably designed receivers.

7.2.2 Television broadcasting systems

Colour television systems are used for virtually all television services and the transmissions are compatible. In television this means that they can be received on a monochrome (black-and-white) receiver and also that they occupy the same bandwidth as a monochrome transmission with the same number of lines and number of pictures per second. All compatible colour television transmissions use the same basic system in which a monochrome signal is transmitted together with a signal giving the colour information. These are known, respectively, as the 'luminance' and 'chrominance' signals.

The initial colour signal generated by the programme source (a camera, video-tape machine or telecine) comprises R (red), G (green) and B (blue) components with amplitudes that, when combined, will give a picture with the correct colours and the correct saturation and luminance. By saturation is meant the amount by which a colour is diluted by white light (in common terms, how intense the colour is) and by luminance is meant the brightness. These R, G and B components are combined to provide:

1. The luminance signal by weighting the R, G and B components and adding them so that the resulting voltage corresponds to the brightness. This signal gives an acceptable picture on a monochrome receiver.
2. A sub-carrier signal (the chrominance signal) whose phase gives the hue of the colour information in the picture and whose amplitude gives the saturation. This signal, therefore, does not contain any luminance (brightness) information. The sub-carrier lies in the video band (the band containing the luminance signal) and its frequency is chosen to minimize interference with the luminance signal.

Figure 7.4 shows the spectrum of the complete video signal for the British colour system – Standard I – (typical of European practice) and it will be seen that the chrominance bandwidth is less than the luminance bandwidth and that the total bandwidth is limited to that of a monochrome transmission. This is possible because the human eye determines the definition (sharpness) of the picture from

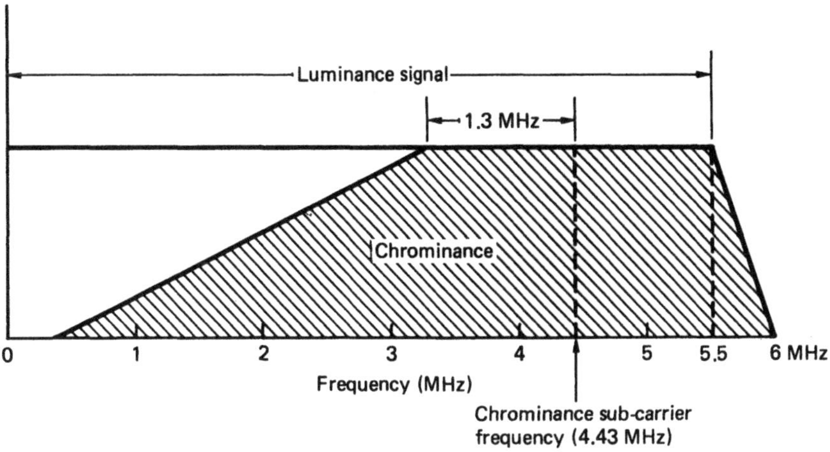

Figure 7.4 Video spectrum of 625-line 50 Hz colour system

the luminance (brightness) rather than from the chrominance (hue and saturation). In the colour receiver, a decoder processes the luminance and chrominance signals to recover the original R, G and B signals which are then fed to the three guns of the colour display tube.

There are three colour television systems in common use that are based on the separate luminance and chrominance principle.

(a) NTSC

The NTSC (National Television System Committee) system was the first compatible colour television to come into use and was developed in the USA. It is used throughout North America on the 525-line 60 fields/second standard. The complete luminance plus chrominance signal is transmitted on each line of the picture and errors in colour sub-carrier phase can result in colour errors in the received picture.

(b) PAL

The PAL (phase alternating line) system was developed in West Germany and is in wide use in Europe on the 625-line 50 fields/second system. PAL is a development of the NTSC system which corrects, automatically, errors in the phase of the colour sub-carrier arising in transmission. The system therefore does not suffer from the colour errors in the viewer's colour picture that are possible with NTSC. The automatic correction is obtained by reversing the sense of phase errors on alternate lines of the picture so that they can be cancelled in the receiver. With the improvement in transmission equipment performance in recent years this improvement in the performance of PAL over NTSC has become of less importance.

(c) SECAM

The SECAM (sequential colour with memory) system was developed in France, where it is currently in use. Like PAL, SECAM is a development of the NTSC system which eradicates hue errors due to phase shifts occurring during transmission. This is achieved by transmitting different components of the colour sub-carrier on alternate lines of the picture. SECAM is less widely used than NTSC and PAL.

All three systems are designed to permit the use of 'inter-carrier' detection of the FM sound transmissions in receivers (hence 100% modulation of the vision carrier is not permitted) and to reduce flicker by interlacing sequentially scanned 'half vertical definition' pictures.

(d) Teletext

This is a system developed jointly by the BBC and National Transcommunications Ltd (previously the IBA) by which digital information is added to standard television transmitters so that viewers with appropriate receivers can display pages of text or graphics as an alternative to pictures. Typically one page of text is an index for the others, which often include headline news, weather forecasts, financial news, advertisements and other information. The digital information is transmitted during that latter part of the frame blanking periods so that it is not visible when normal pictures are being viewed.

(e) Stereophonic sound

In the UK high-quality stereophonic sound is added to television transmissions to provide an alternative to the standard monophonic FM transmissions. A system of near-instantaneous companded audio multiplex – NICAM – is used and the digital information is radiated on a sub-carrier, with an amplitude 20 dB below the peak sync vision carrier, at 6.552 MHz above the vision carrier. Further information about the NICAM system is given in Section 5.6.

7.2.3 Radio frequency (RF) transmission

Television transmissions are more tolerant as regards non-linear distortion than are audio transmissions, and as a result the RF bandwidth can be reduced by using a form of SSB (single-sideband) transmission without requiring additional

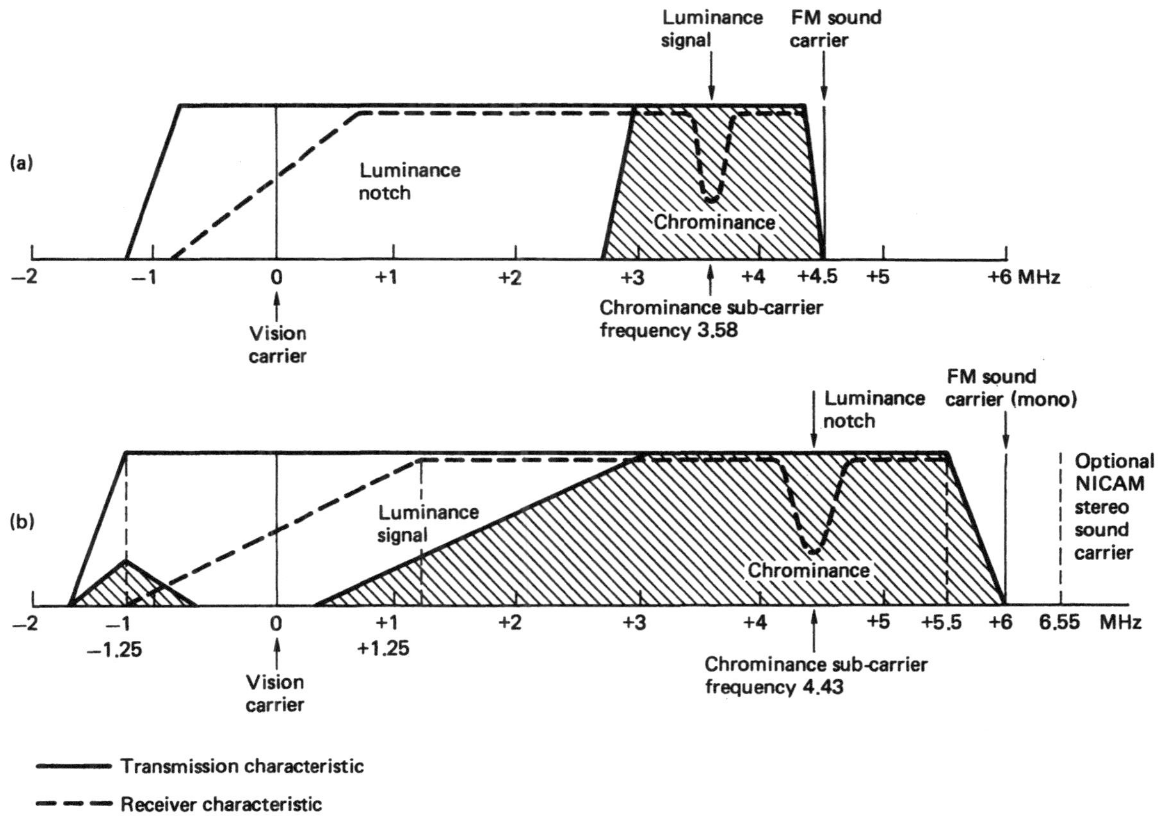

Figure 7.5 Colour television RF spectra. (a) North American (525/60); (b) European (625/50)

complexity in the receiver. The system is known as 'vestigial sideband transmission', in which the upper sideband is transmitted in full but the higher modulating frequencies of the lower sideband are suppressed. The RF spectrums of North American and European transmissions are shown in Figures 7.5(a) and (b). It should be noted that for all colour systems the colour sub-carrier frequency is chosen so that the 'dot pattern' interference it causes in the picture on a monochrome receiver tends to cancel on alternate lines and/or fields. In a colour receiver a 'notch' filter is provided (see Figure 7.5(b)) to remove the dot pattern interference.

In a colour television system each transmitted line (excepting those associated with field synchronization) contains the picture signal (luminance plus chrominance) together with a short burst of colour sub-carrier (to enable demodulation of the chrominance signal in the correct phase) and a synchronizing pulse or pulses (to synchronize the line and field frequency generators in the receiver) see Figure 7.6(a). Virtually all television systems use negative AM in which the maximum carrier amplitude occurs when the synchronizing pulses are transmitted. The minimum carrier amplitude, therefore, occurs when the picture signal is at 'peak-white' with 100% colour bars (see Figure 7.6(b)). As can be seen, the transmitted power varies considerably throughout each line period but, for convenience, the vision power of a television transmitter (using negative modulation) is always quoted as the power occurring at the peaks of the synchronizing pulses (e.g. 10 kW peak-sync). Thus normal maximum and average power, for a black picture and programme, respectively, is 0.63 and 0.44 of this, plus the power of the audio transmitter.

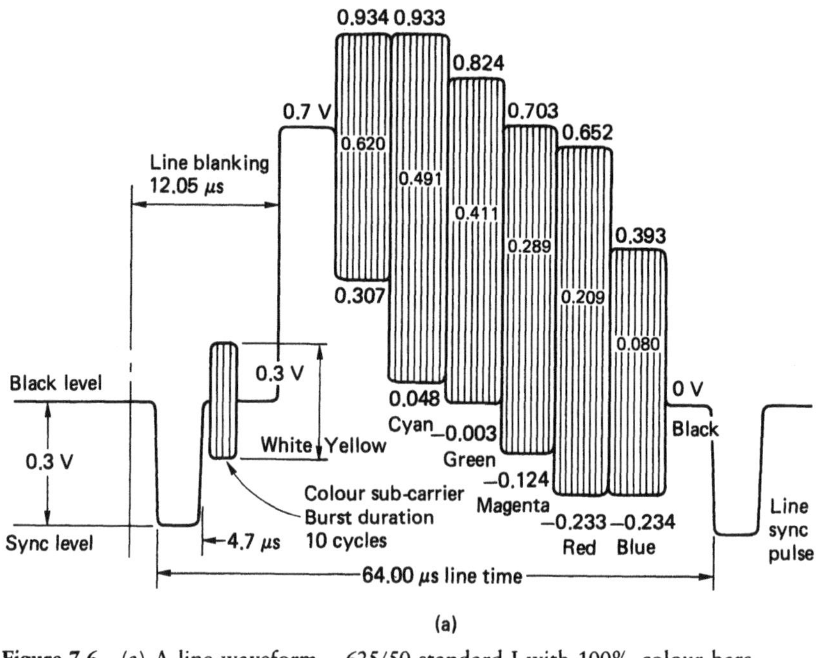

Figure 7.6 (a) A line waveform – 625/50 standard I with 100% colour bars.

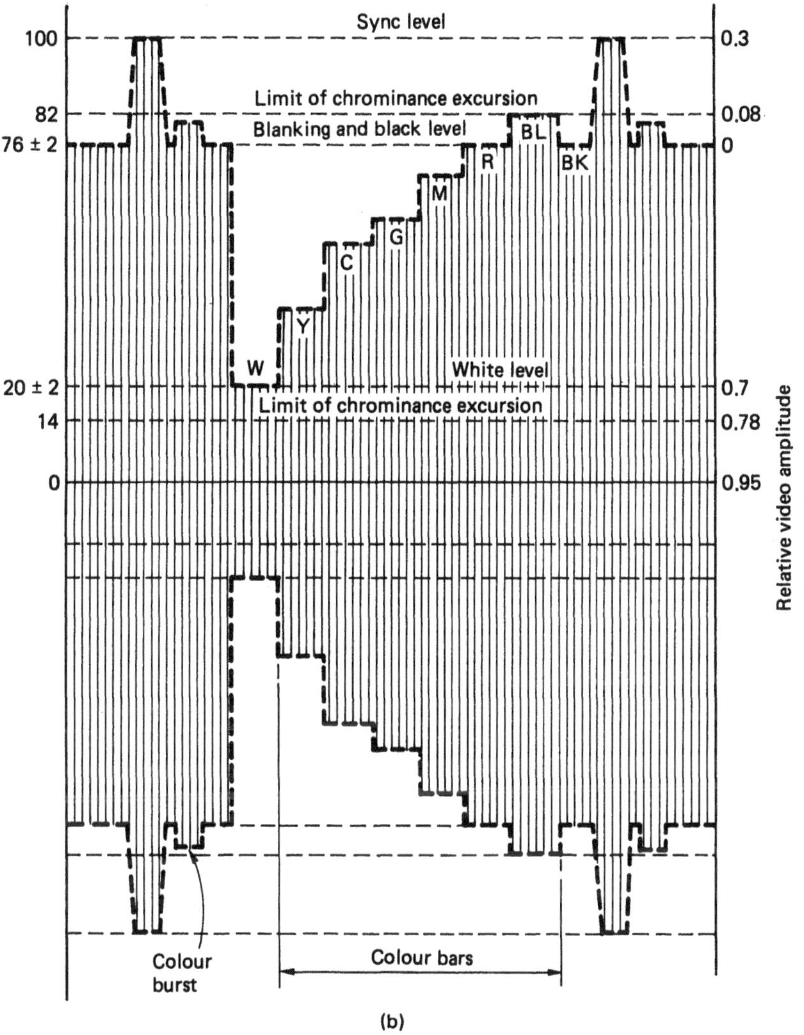

(b) Modulated carrier waveform – 625/50 standard I – as transmitted after VSB shaping: black, white and 100% colour bars. (Note: due to the scale of the drawing, only the peaks of the modulated RF carrier are shown during the periods of the sub-carrier burst and colour bars)

7.3 Thermionic tubes for VHF and UHF applications

7.3.1 Tetrode tubes

The first RF power tubes used in broadcast transmitters were triodes (comprising a cathode, control grid and anode) and had relatively low gain. Virtually all tubes now used for VHF and UHF radio and television applications are tetrodes, which have the advantages of high power rating, higher gain and low grid-anode capacitance (which simplifies transmitter design). The tetrode is cheap, compact and efficient but its design becomes increasingly difficult as the frequency is increased. This is because the transit time of the electrons and the effect of inter-electrode capacitances increase the losses and reduce the gain. These effects can only be reduced by using a small tube structure which, in turn, requires close electrode spacings and makes anode cooling more difficult.

Early designs of VHF and UHF tetrodes had limited gain (up to about 6 dB) and limited anode and grid dissipations. However, the use of the pyrolytic graphite material (referred to in Section 6.3.2) for the grid together with improved methods of construction has made very small and accurate electrode spacings possible and made available tubes with much improved performance. Tetrode gains of around 15 dB are now possible and, together with modern solid-state techniques, single-tube transmitters with output powers of up to 100 kW in the VHF band and over 20 kW in the UHF band are possible.

For television applications it is possible to use a single tetrode for common vision–sound amplification. There is an obvious saving of one complete transmitter chain and the vision–sound diplexer but, against this, the generation of IPs (intermodulation products between vision, colour sub-carrier and sound signals which degrade pictures and may cause out-of-channel interference) means that a higher power capacity tetrode, de-rated, must be used at vision output powers above about 10 kW in the VHF band and 5 kW in the UHF band except as a standby facility in the event of failure of the vision or sound transmitter.

7.3.2 The klystron

The klystron is illustrated in schematic form in Figure 7.7 and takes advantage of the electron transit time that places a limit on the gain and power output of a tetrode. The cathode emits a stream of electrons whose velocity is modulated by the RF waveform applied at the input cavity. As the electrons travel down the drift tube, because of their different velocities they form bunches and arrive at the output cavity (where the RF output power is removed) in the form of a high beam current, amplitude modulated by the RF waveform. The bunching process is assisted by the resonant cavities and the beam energy remaining after the RF output power is removed is dissipated in the collector to which a high voltage is applied.

The klystron has a much higher gain than the tetrode (in excess of 30 dB as opposed to around 15 dB) and this simplifies transmitter design by reducing the drive power required. Current transmitter designs have output powers of over 60 kW with a single vision klystron and all-solid-state drive. The disadvantage of the klystron is its low efficiency which results from the fact that the beam current must be capable of accommodating the peaks of the synchronizing pulses (as previously explained, virtually all television systems use negative modulation). Further, early UHF klystron designs were based on types intended for

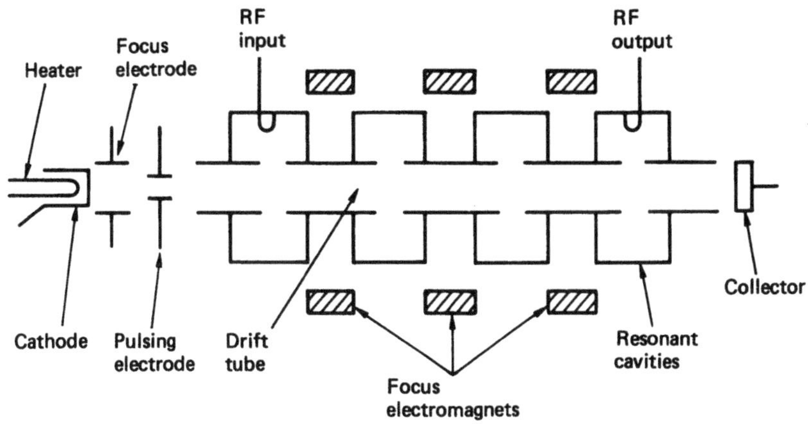

Figure 7.7 Four-cavity klystron

completely different applications (e.g. tropospheric scatter communications links). For these reasons, the efficiency of early klystron vision amplifiers was in the region of only 30% and the overall efficiency of vision–sound-transmitting equipment was around 20%. However, it should be noted that the problem of low efficiency does not apply to a sound klystron amplifier as FM is used and the amplitude of the RF signal is constant. Considerable development work, based on computer modelling, has taken place to develop klystron designs with improved efficiency and wide operating bandwidth. This work has resulted in external cavity klystrons with efficiencies in the region of 45%, an operating bandwidth covering the UHF band (470–860 MHz) and output powers of over 60 kW. The cavities of an integral cavity klystron form part of the tube itself.

An effective means of raising efficiency is the 'pulsed' klystron in which the beam current is controlled by the pulsing electrode shown in Figure 7.7. The beam current is kept at black level (the maximum necessary) during the period of the picture signal but, during the synchronizing period, it is 'pulsed' to the higher value necessary to accommodate the sync pulses. As a result, the mean beam current is reduced and the power lost through dissipation at the collector is reduced. The pulsed klystron technique is now in common use and results in klystron amplifier efficiencies of 65–75%.

7.3.3 High-efficiency klystrons

The efficiency of a pulsed vision klystron, although much higher than that of the conventional klystron, still represents a serious waste of mains power. To increase efficiency still further, and reduce power consumption costs, there are currently three approaches to klystron design and operation. Before discussing these developments, however, it is necessary to define the efficiency of a vision klystron amplifier, taking account of the fact that the mains power consumed varies widely for the sync pulse and picture information periods of the signal. The efficiency is usually defined as the ratio (expressed as a percentage) of the peak-sync output power and the average DC input power and this is known as the figure of merit (FOM). On this basis, an unpulsed klystron has an FOM of around 50% whereas a pulsed klystron has an FOM around 85%. For example, an unpulsed 60-kW (peak-sync) klystron will require a DC input power of 120 kW but, when pulsed, this will be reduced to 71 kW. It is important to note that, defined in this way, the FOM can be greater than 100% but this, of course, does not mean that the average RF output power of the klystron is greater than the average DC input power!

(a) Full-time modulation

An extension of the pulsed-beam klystron technique offers the prospect of a further increase in efficiency. Here the beam current, instead of being pulsed between two values (high to accommodate the sync pulses and low to accommodate the video waveform (the picture signal)), is varied continuously. The aim is to provide the minimum beam current necessary to accommodate the RF signal at each point in its cycle, including sync pulses and video waveform. Attempts have been made to develop this principle (known as full-time modulation) and to modulate the beam so that it follows, continuously, the video waveform. This requires the addition of a suitable electrode to the klystron, probably in the form of a grid. At the time of writing, the continuously pulse-modulated klystron is not in general use, but tests indicate that FOMs well in excess of 100% should be possible.

(b) The multi-stage-depressed collector (MSDC) klystron

The power lost in a klystron is dissipated at the collector and increases with the collector potential – the EHT (extra-high tension) voltage. For many years, therefore, attempts were made to raise the efficiency of a klystron by reducing the operating potential of the collector. This is not as straightforward as it sounds, as it results in a degradation in performance which is due to electrons reflected from the collector (secondary emission) travelling back down the drift tube. This problem has been overcome by Varian with an arrangement that uses a multi-stage collector with a range of EHT voltages. The pulsed MSDC klystron can have an FOM in excess of 120%, which means that a 60-kW (peak-sync) klystron requires an average DC input power of 50 kW compared with 70 kW for a conventional, similarly rated, pulsed klystron.

(c) The klystrode

A quite different approach to the problem of increasing klystron efficiency has been taken by Eimac. Their development work has lead to the 'Class B klystrode', which aims at combining the high efficiency of the tetrode with the high output power and high reliability of the klystron. As illustrated in Figure 7.8, an electron beam is formed at the cathode and amplitude modulated by a grid exactly as in a tetrode. The beam is then accelerated down a drift tube towards the collector and the RF signal energy is removed by a tuned cavity as in a klystron. The grid is biased to cut off so that the beam current depends on the RF signal amplitude at all times and is the minimum necessary to accommodate the

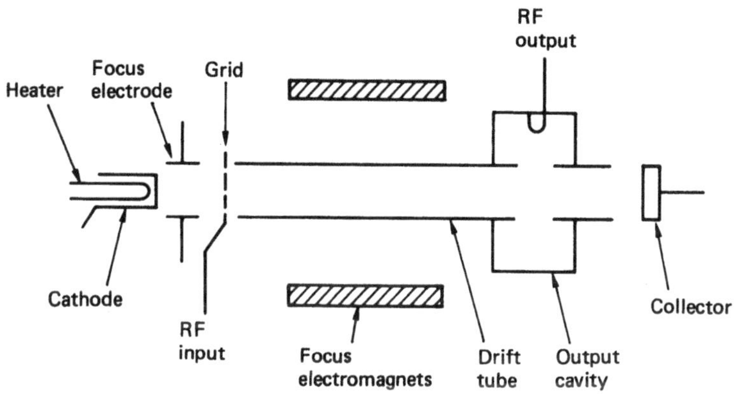

Figure 7.8 Class B klystron

signal. The klystrode, therefore, like a tetrode, operates in the efficient class B mode instead of the inefficient class A mode of a conventional klystron. Unlike the tetrode, however, there is no degradation of performance due to electron transit time. The klystrode FOM is about 120% (similar to that of the MSDC klystron) but the gain is lower (about 20 dB as opposed to 30 dB). However, as RF power outputs of several kilowatts are possible with solid-state driver stages this is not a significant disadvantage. The 'bunching' of the beam occurs at the grid so that a shorter drift tube can be used, and a klystrode is therefore smaller than a corresponding klystron.

(d) Klystron operation

The relationship between the input and output signals of a klystron is not linear, particularly as saturation is approached. By saturation is meant the condition when the beam current is completely bunched so that the minimum current approaches zero (this condition can be referred to as full modulation of the beam). However, the efficiency of a klystron is increased as the modulation of the beam current is increased so that high efficiency can only be achieved at the expense of increased non-linearity. As a result, when operating a klystron in the most efficient mode (with the beam fully modulated) it is necessary to incorporate non-linearity correction in the low-power stages of a transmitter, which is easier to achieve with negative rather than positive modulation.

Although it is theoretically possible to design klystrons for very high powers, the maximum vision power commercially available is in the region of 60-kW–70-kW peak-sync. As has been explained, coverage is limited by the semi-optical mode of transmission in the UHF band and, when using a high-gain antenna, a transmitter power above 70 kW is unlikely to be cost-effective. The size of a klystron increases with decreasing frequency and, even in the UHF bands, the length is, typically, around 1.5 m. For this reason there are no VHF klystrons and tetrodes are used in the output stages of high-power transmitters in these bands.

The fact that the performances of the klystron and klystrode are not limited by electron transit time means that small, and closely spaced, electrodes are not required and, as a result, very long lives are achieved. Klystron lives can be in excess of 30 000 to 40 000 hours and compare very favourably with the figure of less than 10 000 hours common with tetrodes. Further, unlike a tetrode, a klystron is not likely to suffer catastrophic failure and can be replaced at a convenient time when its performance falls below a specified minimum. Klystrons may be of integral cavity type (they form part of the tube) or of external cavity type (they form part of the transmitter and have to be fitted to the tube). The integral cavity type has the advantage of quick installation and minimal adjustment and the external cavity type has the advantage of lower cost and (by tuning) operation over the whole UHF band. The tendency of transmitter designers, at the time of writing, is towards the external cavity type.

7.4 VHF/FM transmitters

A VHF/FM transmitter comprises, essentially, a drive unit followed by an RF amplifier as illustrated in Figure 7.9(a). The drive is, in effect, a low-power transmitter that will generate a complete FM broadcast signal at any frequency in the VHF/FM broadcast band (87.5–108 MHz). The RF output power of the drive is low (typically in the range 3–15 W) so that an all-solid-state design is easily achieved. The drive is followed by a pre-amplifier and an output amplifier giving the required output power. This arrangement has the advantage of standardization, as a single drive design together with a range of pre-amplifiers and power amplifiers will meet all operational requirements. The drive arrangement is illustrated in schematic form in Figure 7.9(b) and it will be seen that complete baseband signal is first generated and then used to frequency modulate the carrier.

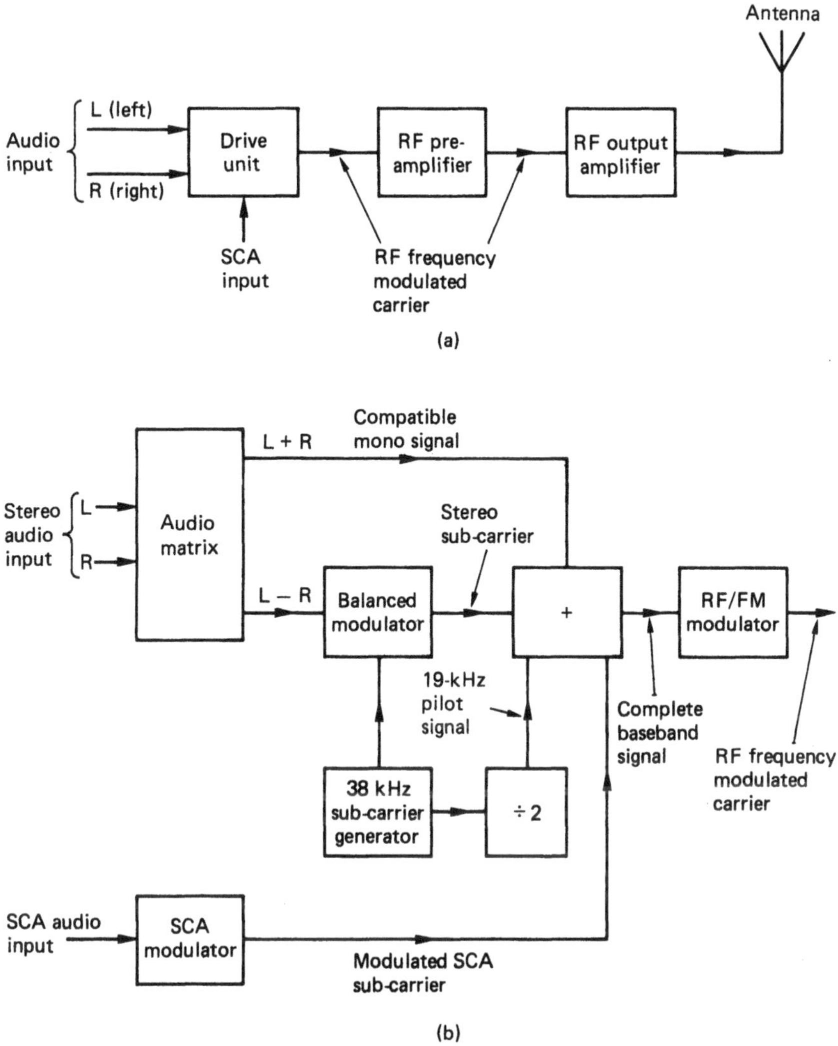

Figure 7.9 VHF/FM transmitters. (a) FM transmitter schematic. (b) Stereo/SCA FM drive unit schematic

7.4.1 Solid-state transmitters and transposers

As in the case of MF transmission there has been a steady increase in the output power available from all-solid-state VHF/FM transmitters and, at the time of writing, powers up to 50 kW are available. In the VHF band, the curvature of the earth and shadowing by natural and man-made obstacles limit the coverage to about 100 km (60 miles), however high the effective radiated power (ERP, the product of transmitter power and antenna gain). As a result, unlike the case in the LF, MF and HF bands, there is no requirement for transmitter powers much in excess of 100 kW (in fact, the transmitter power in most cases does not exceed 50 kW). It is clear, therefore, that the stage had been reached at which all-solid-state transmitters will satisfy most, if not all, VHF/FM transmitter power requirements. As in the lower frequency bands, solid-state transmitters have the advantages of built-in redundancy, reliability, compactness and high efficiency. Further, for FM transmission, there is no linearity requirement (the carrier frequency, not the carrier amplitude, is modulated) and the power amplifier stages can operate in the high-efficiency class C mode so that the output power of each transistor module is maximized. As a result, a single transistor can produce up to 200 W RF output power so that, for example, a 10-kW transmitter can be designed using only 50 transistors in the output stage.

In cases where a VHF/FM station can be fed by another station (RBR reception) the drive in Figure 7.8(a) may be replaced by a unit comprising a receiver and frequency changer and the arrangement is known as a 'transposer'. Transposer stations are normally used to fill gaps in the coverage provided by a network of higher power VHF/FM stations and have output powers ranging from a few watts up to several kilowatts. However, if the received signal is weak and/or the spacing between the received and transmitted channels is small it may be necessary to use a receiver–drive combination to achieve a workable system. Equipments at these power levels are almost invariably of all-solid-state design.

Although all-solid-state transmitters have many advantages, it should be noted that there are some disadvantages. The multiplicity of paths (i.e. the number of transistor modules combined in parallel in the output stage) can lead to phase differences that militate against an acceptable frequency response. Further, the cooling of power transistors requires a larger volume of air than is needed for a tetrode output stage. Finally, power transistors are prone to sudden failure although, in this respect, they cannot be said to be markedly inferior to tetrodes. However, these shortcomings cannot be said to outweigh the advantages of all-solid-state designs and, in any event, are being overcome. It seems, therefore, that the use of all-solid-state VHF/FM transmitters will continue to expand.

7.4.2 Tetrode transmitters

Transmitters with tubed output stages almost invariably comprise a drive unit followed (where necessary) by a solid-state amplifier and a single tetrode. As in the case of transistor output stages, linear amplification is not required and the tetrode can be operated in class B or class C to ensure high efficiency. When compared with all-solid-state equipment, the tetrode output stage transmitters have a number of advantages:

1. The overall transmitter arrangement is more compact.

2. The power supply unit, although high-voltage, is smaller.
3. The overall transmitter efficiency is at least equal and can be higher.

The disadvantages are that tetrode lives are relatively short (at best 5000 to 8000 hours) so that the operating reliability is lower. As a result, to equal the reliability of a solid-state transmitter, it may be necessary to provide a second transmitter as a standby.

7.4.3 Combining networks

It is common in VHF/FM broadcasting to radiate more than one programme from the same transmitting site and it is, of course, possible to do this by using a separate antenna for each programme (though band-pass filters may be required to reduce the coupling between transmitters and hence the radiation of intermodulation products). It is, however, far more economical and preferable to achieve equality of service areas to use a single antenna system and to combine the separate programmes at ground level. This approach requires a combining network satisfying three conditions:

1. There must be low insertion loss between each transmitter and the antenna to minimize reduction of the transmitted signal and any cooling requirements.
2. There must be adequate isolation between the transmitters to avoid the generation of intermodulation products causing interference to other transmissions – including RBL (re-broadcast link) feeds, if used – or transmitter malfunction.
3. The impedance at each transmitter input must result in a sufficiently low voltage-standing-wave ratio (VSWR) to avoid distortion of the transmitted signal.

The simplest design of combining network meeting these requirements is the 'star-connected' arrangement illustrated in Figure 7.10(a) for the case of three transmissions. Here the resonators associated with each transmitter are arranged to pass energy at the frequency of that transmitter (the pass frequency) and reject energy at the frequencies of the other two transmitters (the stop frequencies). The two resonators are spaced apart by an electrical distance such that the mismatches they produce at the pass frequency tend to cancel. At the same time, the two resonators reinforce each other and produce added loss at the stop frequencies.

The disadvantage of a star-connected filter is that it is difficult to achieve adequate isolation between transmitters and a low VSWR when the frequency spacing between transmissions is less than 1.5–2.0 MHz or several transmitters are so connected. For smaller frequency spacings hybrid directional coupler or 'ring' filters can be used: the arrangement of the latter is shown in Figure 7.10(b), and of the former in Figure 7.15. The resistive load absorbs energy reflected back

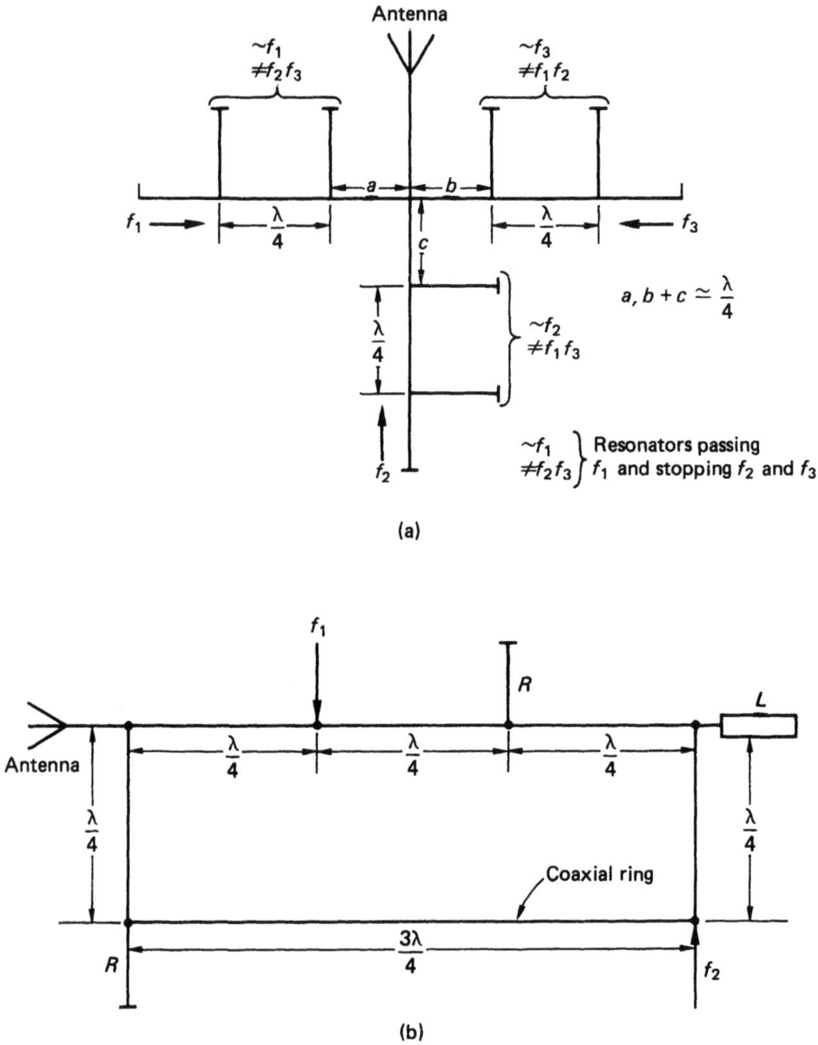

Figure 7.10 VHF/FM transmitter combining filters. (a) Three-transmitter 'star-connected' combining unit. (b) Coaxial-ring combining unit. f_1 and f_2, transmitter carrier frequencies; L, resistive load; R, resonators with zero impedance at f_1 and infinite impedance at f_2

towards the antenna due to mismatch and results in a well-matched input impedance and high isolation between transmitters. The drawback is that one ring or hybrid directional coupler filter has only two inputs and $n-1$ filters are therefore required to combine n transmissions (e.g. two ring or hybrid directional coupler filters are required to combine three transmissions, etc.).

Figure 7.11 Four VHF/FM radio combining units, with monitoring and protection equipment, combining pairs of 10-kW transmitters for each of three programmes, with the capacity to extend to five programmes by the addition of four more units (Courtesy of BBC London)

7.5 Television transmitters

Television transmitters are inevitably more complex than those designed for VHF/FM applications and, when medium or high powers are required, comprise two transmitters, one for vision and one for sound, as illustrated in Figure 7.12. The sound signal is frequency modulated and usually has a carrier power of one-tenth of the peak-sync vision power. Television transmitters are designed for the VHF Band III (170–250 MHz), the UHF Bands IV and V (470–860 MHz) and, in a few cases, for the VHF Band I (47–68 MHz). Until relatively recently, except at very low powers, the output stages of a VHF or low-power UHF transmitter used a vacuum tube (a tetrode or grounded-grid triode) and those of medium–high-power UHF transmitters used klystrons. Further, for high-power transmitters it was sometimes necessary to drive the output stage with a vacuum tube amplifier. However, RF powers at the kilowatt level and above are now being achieved with solid-state output stages and, almost without exception, a tubed television sound or vision transmitter uses only one tetrode or klystron in the output stage.

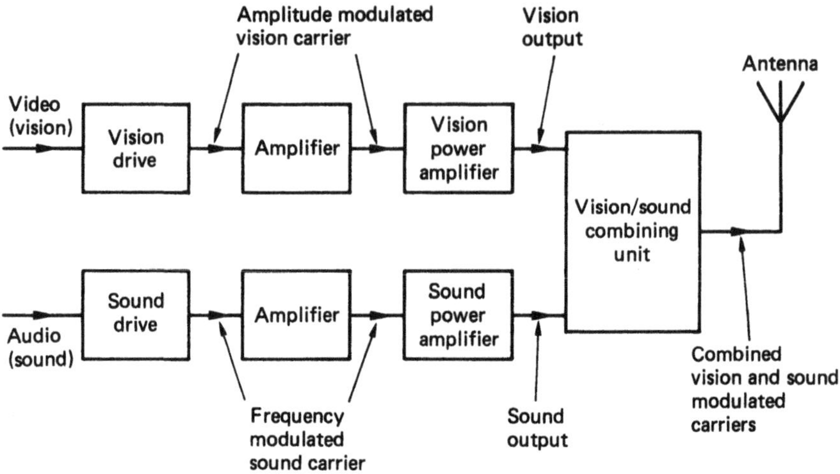

Figure 7.12 Television transmitter with separate amplification

Standardization of the units comprising television transmitters which have different output powers, and which operate in different frequency bands, reduces manufacturing costs and simplifies maintenance procedures. An important measure of standardization can be achieved with intermediate frequency (IF) modulation in which an RF signal is generated with the same carrier frequency (around 40 MHz) irrespective of the actual broadcast carrier frequency. This permits the use of a standard drive unit which, in conjunction with suitable interfacing amplifiers, can be used in conjunction with klystron, tetrode and solid-state output stages operating in Bands, I, III, IV and V. An IF drive gives standard vision and sound output levels and includes vision linearity correction circuits (mainly for klystron operation) and heterodyne oscillators to change the intermediate frequencies to the required vision and sound carrier frequencies. An output providing a drive for a klystron pulsing electrode is also provided. Drives (usually referred to as 'exciters' in North America) usually allow for the transmission of a second sound channel in accordance with internationally agreed 'second programme standards'. This second channel can, of course, be used for stereographic sound, which in the UK is transmitted with a digital system.

In North America, the television coverage problem differs significantly from that in Europe where, generally speaking, countries are densely populated and a national television system has to be planned to achieve an area coverage approaching 100%. The geographical area of North America is very large, much of the population is concentrated in towns and cities, and there are large areas that are relatively sparsely populated. As a result, particularly in the USA, many transmitting stations are designed to cover a single town, or city, rather than to give area coverage as is common in Europe.

The importance of city coverage in North America is reflected in the range of equipment available, where the emphasis is on medium- and high-power transmitters and there are fewer manufacturers of the low-power transmitters and transposers (see Figure 7.13(a)) that form a large part of the European market. In Europe, most television transmitters are sited in rural areas and provide coverage of a fairly wide area that includes one, or more, towns or cities. In this case a wide-area coverage can be obtained with a high-gain antenna and a transmitter power up to 60 kW. However, in North America, television transmitters are often sited in, or adjacent to, a city on a high structure or on the top of a high building. In this case, a high-gain antenna cannot be used, as the nulls in the vertical radiation pattern (VRP) degrade reception in the neighbourhood of the station. Thus, particularly in the case of UHF transmission, a low-gain antenna with broad-lobe VRP must be used with a correspondingly increased transmitter power. High transmitter powers are also used, in conjunction with very high masts, to provide coverage of highly populated areas in flat terrain. This need for increased transmitter power has led to the development of VHF transmitters with output powers up to 100 kW and UHF transmitter combinations with output powers up to 240 kW. When these high powers are used, the high cost of mains electricity makes the conversion efficiency from mains input power to RF peak-sync output power increasingly important.

7.5.1 Solid-state transmitters and transposers

A solid-state vision transmitter or transposer output stage for operation in Band III, IV or V poses a more difficult design problem than that encountered in the VHF/FM Band II case. Not only is the carrier frequency higher but also linear amplification is required and it is not possible to operate the output stage in the efficient class C mode. Despite this, however, there is a continuing and steady increase in the power available from all-solid-state VHF and UHF transmitters and, at the time of writing, vision output powers up to 60 kW are available in Band III. The generation of high powers becomes increasingly difficult with increasing frequency and, in the UHF bands, all-solid-state vision powers are limited to about 10 kW. However, the advantages of the all-solid-state vision transmitter as regards reliability, safety (due to the low supply voltage) and overall efficiency (similar to that of a tetrode) are such that development work aimed at higher powers in the UHF bands is bound to continue. It must be said, however, that all-solid-state UHF vision transmitters for the highest powers (at the 60-kW level and above) are not likely to become available for many years to come.

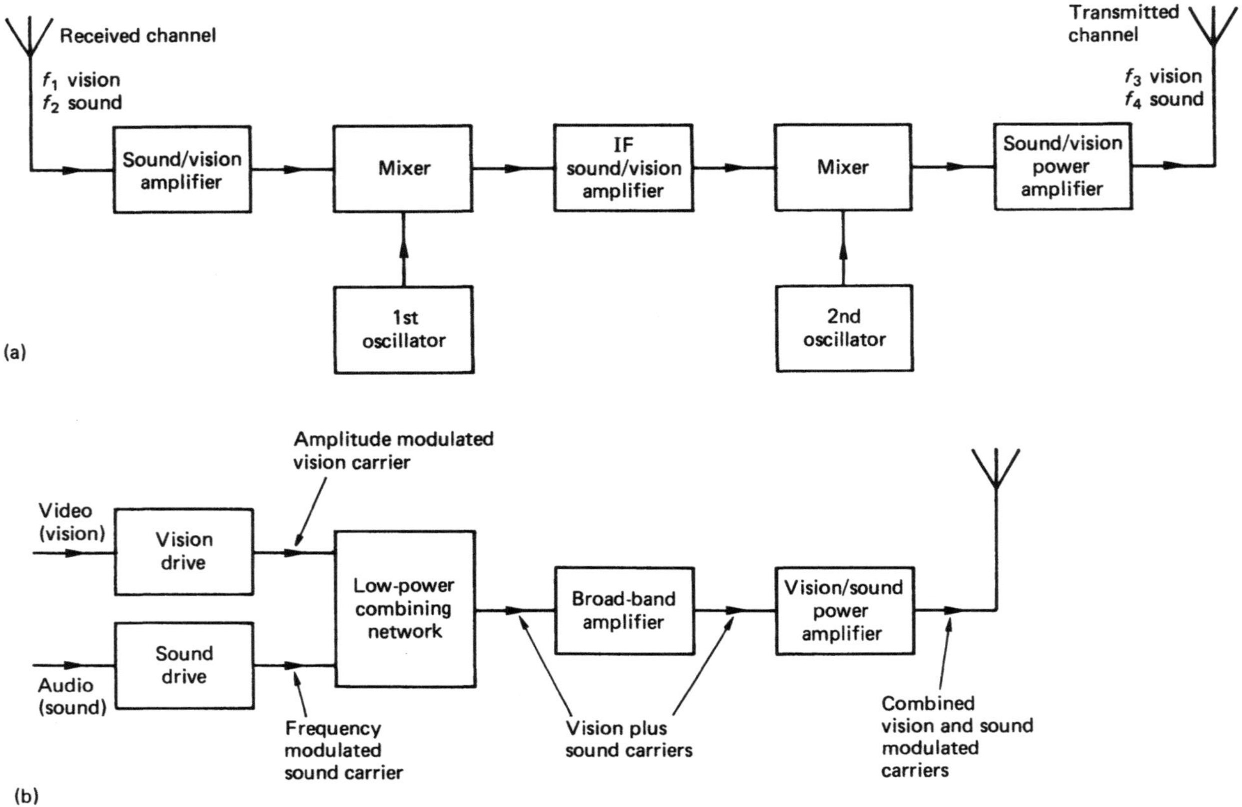

Figure 7.13 (a) Television transposer (combined processing and amplification). (b) Television transmitter with combined amplification

7.5.2 Tetrode transmitters

The size of a klystron becomes too great for its use in the VHF band, so that the only available thermionic tube is the tetrode. At these frequencies the effects of inter-electrode capacitance and electron transit time make it necessary to use a small tube structure and this leads inevitably to tube-cooling problems. As a result, a UHF tetrode will produce significantly less power than one operated in the VHF band. Nevertheless the advantages offered by the tetrode output stage as regards compactness and cheapness are such that single tetrode vision transmitters are available for the highest powers used for VHF transmission (over 50 kW) and for the low to medium powers used for UHF transmission (up to 20 kW).

The power of the sound component of a television transmission is lower than the vision power and does not require linear amplification. This makes the combination of a tetrode vision output stage with a solid-state sound output stage attractive. This approach is more common in the VHF band where it is easier to achieve high powers with solid-state circuits.

7.5.3 Klystron transmitters

The disadvantages of the tetrode as regards output power, cooling, gain and limited life in the UHF band limit its use to a maximum peak-sync vision power of between 10 and 20 kW. For higher powers, the klystron offers many advantages and is in common use. The disadvantage of low efficiency has been largely overcome with the introduction of the beam-pulsing technique, and the development of the MSDC klystron and the klystrode. The maximum peak-sync vision power available from a single klystron (or klystrode) is 70 kW and, for higher powers, one or more klystrons (or klystrodes) are operated in parallel. The highest vision power normally available (mostly used in conjunction with a low-gain antenna for city coverage in North America) is 240 kW. A 240-kW transmitter typically uses four klystrons (or klystrodes) in parallel in the vision output stage and two in parallel in the sound output stage. Also available is a 120-kW arrangement using three klystrons (two in parallel in the vision output stage and one in the sound output stage).

7.5.4 Vision–sound combining arrangements

It would be very uneconomic to use separate vision and sound antennas and, invariably, the vision and sound transmissions are combined and radiated from a common antenna. It might be thought that universally the most economic method of achieving this combination is to combine the vision and sound signals at a low power level and to use a common output stage as illustrated in Figure 7.13(b). This system (known as mono or combined amplification) results in a considerable reduction in transmitter size and cost. At a low power level a relatively simple combining network can be used since the insertion loss introduced is unimportant and can be compensated by an increase in gain in the driver stages. The penalty paid for the simplicity of combined amplification is that the power output from the output stage must be reduced to eliminate the generation of intermodulation products. With combined amplification, the maximum peak-sync power of a tetrode, klystron or klystrode is reduced to about 50% of that obtainable when used as a 'vision only' amplifier. As a result, on efficiency and operating cost grounds, combined amplification can only be used at low and medium powers in the VHF and UHF bands.

For the higher radiated powers (over 10 kW peak-sync) it is necessary to combine the outputs of the vision and sound transmitters at the full output power level as shown in Figure 7.12 and any power loss introduced by the filter becomes important and must be kept to a minimum. In the VHF band, the 'ring' type of combining unit comprising coaxial elements and illustrated in Figure 7.10 can be used with the resonators adjusted to accommodate the different bandwidths of the vision and sound transmissions. In the UHF band, the problem of combining unit design is more difficult as the small wavelength makes it difficult to use coaxial elements of a sufficiently large diameter to carry the powers involved and achieve a small insertion loss. Many designs therefore use 'waveguide' resonators which can be physically large and the Marconi 'Rotamode' filter illustrated in Figure 7.14 or the hybrid directional coupler filter illustrated in Figures 7.11 and 7.15 are typical examples.

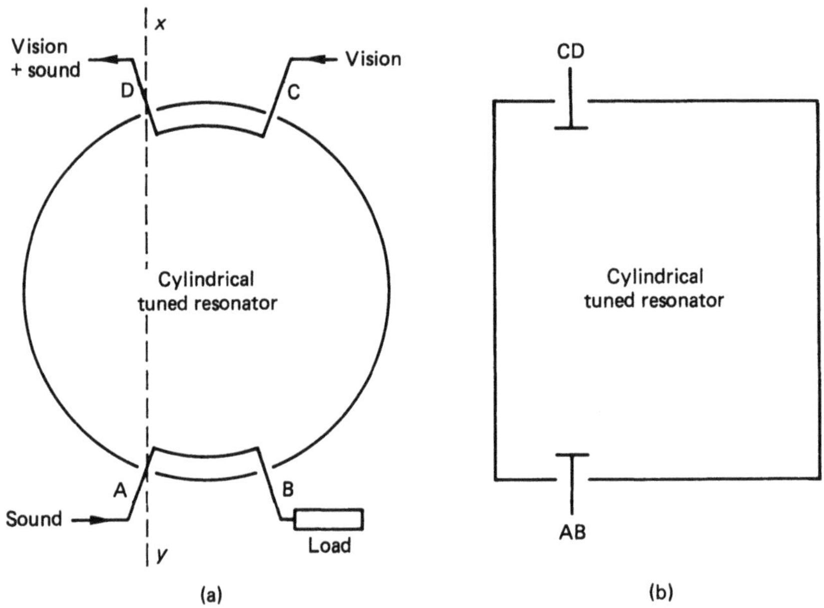

Figure 7.14 Marconi 'Rotamode' vision–sound combining unit. (a) End view. (b) Section on *xy*

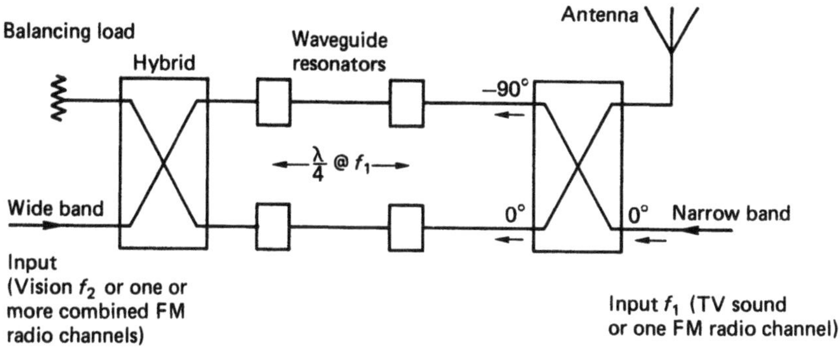

Figure 7.15 Hybrid Directional Coupler Combining Unit: resonators are tuned to reflect f, and pass wideband input

7.5.5 Output-stage cooling

Except for very low output powers all transmitter output stages require cooling to remove excess heat. The simplest and cheapest form of cooling is by air-blast, which avoids the complication of liquid coolants, pumps and heat exchangers but is unsuited to the removal of the large amounts of heat generated by klystrons operating at the 60-kW level. The situation as regards output-stage cooling can be summarized as follows:

1. Solid-state output stages are always air-cooled, since the heat to be removed is spread out among a large number of transistors operated in parallel. The problem here is that transistors do not operate at as high a temperature as thermionic devices and it is necessary to remove heat at a temperature that is relatively low with regard to ambient temperature. This means that a large volume of air is required with a consequently large cooling system.
2. Tetrode output stages are normally air-cooled, as the power levels at which they operate are much lower than those common in LF and MF transmitters and do not require the vapour-phase arrangement described in Chapter 6.
3. Klystrons are used at high powers in the UHF band and normally use vapour-phase cooling. However, the introduction of higher efficiency klystrons and the klystrode results in a reduction of the waste heat to be removed and air-cooling is becoming practicable at even the 70-kW power level. A disadvantage of air-cooling is the level of acoustic noise, which can give rise to unhealthy environments if several transmitters are operated in the same room, in addition to filter maintenance requirements.

7.6 Service planning

If there is a long-term desire to provide complete or virtually complete national coverage with one or more services this should be declared at the time of any enquiry for the first station, as a plan to achieve this should be evolved by the engineering planners before the first station is planned in detail. The CCIR

(International Radio Consultative Committee) have produced theoretical 'grids' of transmitters to show how such coverage could be achieved in the various bands with acceptable levels of co- and adjacent-channel interference for a smooth (spherical) earth. In practice the earth is not 'flat' and for a variety of reasons deviations from the ideal grid of transmitter locations are necessary. Furthermore, when a chain of stations radiating more than one TV or one VHF/FM programme is planned, receiver design places restrictions on the groups of channels which may be allocated to individual and adjacent stations, e.g. stations with contiguous service areas should not use adjacent channels. In the case of VHF/FM services, when more than two programmes are required, it is desirable that all the transmissions on a particular site are separated in frequency by the same interval. By so doing any intermodulation products are generated with the same frequency interval and thus coincide with each other or the carrier frequencies. In this way the number of spurious radiations is reduced and the feasibility of using RBL feeds is considerably enhanced and the risk of interference to out-of-band services minimized. A further constraint on VHF/FM transmissions covering the same area, whether originating from the same or different sites, is that the frequency difference between any two transmissions should not lie within the IF band of standard VHF/FM receivers, i.e. 10.7 ± 0.1 MHz. In order to provide virtually 100% area coverage of a country, Table 7.1 gives the approximate number of channels required in each broadcasting band to provide coverage for one programme.

Table 7.1

Service	*Band*	*Polarization*	*No. of channels*
TV	VHF band I	VP and HP	5*
TV	VHF band II	VP and HP	8*
TV	UHF bands IV and V	VP and HP	11
VHF/FM	VHF band II	VP or SP or CP	14

*The bands allocated to television broadcasting are insufficient for five and eight 625-line channels respectively but were adequate for the now obsolete 405-line system used in the UK.

The number of required channels tends to be the same irrespective of the size or terrain of the country (above a certain minimum size) as the higher frequencies have shorter ranges and suffer more from diffraction losses.

Summarizing, effective use of the limited number of channels available for TV and VHF/FM services presents considerable difficulties, and without the employment of consulting engineers is likely to be unsatisfactory, unless there is extensive in-house expertise. The national broadcasting organizations of the western European countries generally have such experience and it is probably true to say that the BBC has made the most effective use of the available channels in the UK.

Chapter 8

Civil and structural engineering

8.1 Introduction

The starting point for a design of a broadcast transmitting station is a specification of the areas that are to be served, the frequency band to be used and, if available, the operating frequency together with any provision for future services. The station designer must then prepare a costed proposal for a transmitting installation that has an electrical performance (in particular as regards transmitter power and antenna radiation pattern) that will provide, as closely as is practicable, the required broadcast coverage. The station design, in defining the electrical characteristics of the installation, must take account of administrative, local, civil and structural engineering considerations and this may impose constraints on the electrical design. For example, the optimum mast height for an MF antenna may be difficult to accommodate on the chosen site, which may not be located to give optimum coverage. In general, therefore, the overall station design must be a compromise between the coverage and electrical requirements, constraints imposed by local, civil engineering and structural considerations, and, of course, cost.

8.2 Civil engineering

8.2.1 Site selection

When a site for a broadcast transmitting station is selected, a number of considerations must be borne in mind:

1. For domestic broadcasting, a transmitting site at (or near) the centre of the area to be served (which may be a single town or many towns spread over a wide area) will normally result in the most economic arrangement in terms of transmitter power. However, medium- and high-power transmitters are best sited, for reasons described later in this section and Chapter 4, some distance from areas of high population density. For long-distance HF broadcasting (broadcasting to other countries) the siting of a transmitting station is less critical and can therefore be determined largely on grounds of practical convenience.
2. In all frequency bands a site with natural and man-made obstructions in its vicinity should be avoided. Buildings and hills within a few wavelengths of a

site can cause serious distortion of the radiation pattern and in the VHF and UHF bands can generate echoes. Transmitting site terrain characteristics are particularly important at HF stations which broadcast to distant target service areas. In this case the land should be as flat as possible and there should be no obstructions within 30 wavelengths (at the lowest operating frequency) of each antenna, nor should one antenna 'fire' through another. As a rough guide, there should be no obstructions exceeding 20 m within 1–2 km in the direction of transmission of an antenna. For VHF and UHF transmission a site with the greatest possible height over the surrounding terrain should be chosen and in general no other structures of comparable height to the transmitting antennas should be on or within a considerable distance of the site.

3. For MF transmitting sites the local ground conductivity should be as high as possible, as this will minimize ground losses and maximize antenna radiation efficiency. Sites in rocky, dry or sandy terrain should therefore be avoided as far as possible.
4. Although many stations are planned on the basis of calculated coverage it must be stressed that it is difficult to predict coverage accurately. It is therefore advisable (particularly for VHF and UHF transmissions, where accurate prediction is difficult and very laborious) to carry out field-strength survey measurements with low-power test transmissions using the proposed operating frequency, and a balloon to carry the test transmitting antenna if the antenna support structure is not available.
5. Radiation may be a hazard to the public, industrial plant and products. The risk to staff engaged to operate and maintain transmitting equipment can be controlled by training and safety procedures, but for the public at large, restricting access to safe distances is the only practical way. In practice, this is only likely to be necessary at high-power HF stations or radar installations which radiate at low angles of elevation. MF transmissions give rise to a different type of risk. Large metal structures with open loops, e.g. pipework at oil refineries, can act as antennas which pick up a significant amount of power at distances of several kilometres from high-power transmitters and there is a risk of ignition of flammable liquids or gases if sparking occurs. Cross-bonding of such loops can eliminate the risks, but new situations may arise so that it is prudent to site high-power transmitters away from such industrial plants, including explosive and detonator stores.
6. Interference with electronic equipment may be of two distinct types. Field strengths of transmissions which are relatively high (a few volts/metre), but not dangerous to people, may overload electronic equipment. Thus a new relatively low-power transmitter in the centre of a town may so overload the 'front end' of receivers as to render the quality of reception of other services from more distant transmitters unacceptable or impossible, and may interfere with police, ambulances or other public services, which often operate with very low field strengths. Other cases of similar interference with potentially dangerous consequences have been reported, for example, with heart pacemakers and missile detonation systems. The other main type

of interference arises due to the generation and radiation of intermodulation products as a result of unwanted coupling between transmitters on the same site, either via antennas which are in close proximity, combining equipment when transmitters share the same antenna or inadequate screening of transmitter drives. These products, which often fall outside the bands allocated to broadcasting, can be reduced with filters and better screening, but it is sometimes difficult to achieve the required attenuation. Of practical concern are intermodulation products between VHF/FM transmissions which fall in the aircraft navigation and mobile communication bands.

Thus there are good reasons for siting transmitters some distance away from populated areas and airports to minimize the risk of potentially serious consequences.

8.2.2 Power supplies

Most broadcasting stations receive their mains power via underground or open-wire overhead lines. In the industrialized West mains supplies are highly reliable and it is often possible to arrange alternative power supplies from different parts of the power distribution grid so that adequate protection against mains failure is provided. In cases where a second mains power supply is not available a single mains feed can be provided with an automatically controlled standby diesel-alternator for use in the event of a failure of the mains feed. In many Third World countries, broadcasting stations receive their mains power via overhead lines that are often routed over hills and are subject to wind damage and lightning strikes. As a result, the reliability of the mains supply in Third World countries may not be high and may also be subject to voltage variations and surges. The combined effects of lightning strikes and voltage surges can result in damage to low-power transistor stages in the transmitter and, in this case, it is essential that voltage regulation and surge suppression equipment should be provided at the transmitting station. The scale of a reserve power supply will depend on the mains reliability and, if this is below average, it may be necessary to provide a standby power arrangement comprising dual diesel-alternators so that the station can operate for long periods in the absence of the mains supply, possibly at reduced power or half-power if parallel transmitters or reserve equipment is available.

8.2.3 Air conditioning and cooling

In many Third World countries, temperatures are very much higher than in the industrialized West and the build-up of heat inside equipment can lead to the failure of components. It is therefore essential that cooling air is passed through the equipment at the rate specified by the transmitter manufacturer for the maximum ambient temperature at the site or within the building, if the cooling air is not drawn directly from outside the building. This problem is simplified if the transmitter building is equipped with air-conditioning, which may well be essential from the staff point of view in tropical and sub-tropical locations. For unmanned (automatically controlled) stations it may not be economic to provide air-conditioning and adequate equipment cooling is of paramount importance.

Rooms containing free-standing unattended self-ventilated equipment must be adequately ventilated. A typical arrangement of high reliability might consist of a pair of independently thermostatically controlled fans, each capable of providing the necessary cooling, pressurizing the room via air filters, but even with such an arrangement it would be prudent to arrange for the equipment to be fitted with automatic shutdown devices to cover the possibility of damage resulting from overheating in the event of complete failure of the ventilation system.

8.2.4 Buildings, services and site facilities

If capital and maintenance costs are to be minimized, a site must have essential services readily available. These services are mains power, programme feed, road access, water and sewerage; the most important are road access, power supply and programme feed since the need for water and sewerage will depend on whether or not it is to be manned. For very low power stations photovoltaic solar panels, nuclear-powered thermoelectric generators and foot access may be feasible. When a site is sought for a new station, therefore, it is advisable that the broadcaster issues a questionnaire to guide the engineers responsible for site finding, and Figure 8.1 shows an example of a suitable format. It cannot be overstressed that accurate site information is essential as a basis for field-strength prediction and station planning.

Building costs are conveniently estimated on the basis of the cost per square metre of floor area. However, the cost per square metre varies considerably from country to country and it is best to base an estimate on local costs. In the absence of a local figure, US $1200 per square metre (including wiring, plumbing and air-conditioning) for a 'utility' type of building is a useful guide price. This figure is at 1992 prices and is an average for the different types of accommodation (transmitter hall, offices, stores, etc.) and when estimating it is, of course, necessary to make allowance for price inflation. It is worth noting that the cost of utility (industrial-type) buildings is normally only a fraction of the cost of a purpose-built building. Thus, unless there are overriding prestige or environmental reasons for a purpose-built building a utility building offers considerable economy.

In tropical countries, considerable damage can result from vermin which enter the transmitter building through open doors and windows or unsealed holes in the building fabric. Rats can eat cables, lizards can short-circuit

SITE SURVEY QUESTIONNAIRE

Project Number and Title: (e.g. 1234/Zambia MF)..............................

1.	SITE LOCATION	
1.1	Name of site (if relevant)	....
1.2	Map reference	....
1.3	Site owner	....
1.4	Name and distance of nearest town	....
2.	TRANSMISSION INFORMATION	
2.1	Frequency band (LF/MF/HF/VHF/UHF)	....
2.2	Operating frequency (if known)	....
2.3	Transmitter power (if known)	....
2.4	Antenna type (omni or directional)	....
3.	SITE CHARACTERISTICS	
3.1	Site area (with sketch site plan)	....
3.2	Site terrain (rock/sandy/arable)	....
3.3	Climate: dry/wet, max./min. temps	....
3.4	Obstructions in site vicinity	....
3.5	Ground slope over site	....
3.6	Site height above mean sea level	....
3.7	Site height above mean terrain height	....
4.	AVAILABILITY OF SERVICES	
4.1	Distance to nearest water supply	....
4.2	Distance to programme feed	....
4.3	Distance to mains power supply	....
4.4	Power supply type (e.g. 415/3-phase)	....
4.5	Power supply reliability (good/poor)	....
4.6	Distance to nearest road access	....
4.7	Distance to main sewerage	....

Figure 8.1 Specimen site survey questionnaire

high-voltage rectifiers, and ants and spiders can enter the transmitter enclosure, adding to the risk of flashover. An air-conditioned building with sealed windows and self-closing double doors will provide a high degree of protection against this hazard. In addition, holes should be sealed against rats with concrete or with metal plates or steel wool.

The site facilities that must be allowed for when planning a station and their approximate costs are:

Drainage	10% of building costs
Fencing	US $35 per metre
Hard standing (car parking, etc.)	US $55 per sq. metre
Access road (3 m wide)	US $75 per metre

All the costs given are approximate and, since they are at 1992 prices, they must be adjusted for inflation.

8.2.5 Security

There is an increasing incidence of terrorist activity and malicious damage in all countries and it is becoming necessary to take comprehensive security precautions in both the industrialized West and the Third World. However, even if guards are employed (which may well be prohibitively expensive) it is impossible to ensure complete security for a transmitting installation. Further, automatically controlled, unmanned sites are obviously particularly difficult to protect. The following measures, although they will not obviate the risk of damage, will at least reduce the possibility:

1. As much equipment as possible should be accommodated in the transmitter building. External doors giving access to the building should be faced with steel, and windows (if they are essential) should be narrow with 'bandit' (highly toughened) glass. Openings in the building (for feeder and mains supply egress and ingress) should be sealed with steel plates or reinforced concrete.
2. Where possible, feeders from the transmitter to the antennas should be of the coaxial type and should be buried in ducts. The ducts should, if possible, be camouflaged so that their location is not immediately obvious.
3. Masts and tower bases and stay-blocks should be fenced in with high-grade chain-link fences, preferably with a height of 3 m.
4. If possible an intruder alarm system should be installed to switch on sirens and floodlights and alert local police or security forces.
5. Notices which 'threaten' rather than 'deter' should be used, e.g. 'Radiation Hazard'.

These precautions are obviously expensive from both the capital and revenue points of view and the extent to which they are implemented depends on local circumstances. This is a decision that can only be made by the broadcaster who is initiating the project.

8.3 Antenna support structures

8.3.1 Masts and towers

The support structures for broadcast antennas fall into two categories.

Self-supporting towers occupy a small area of ground and are therefore attractive when the available site area is restricted. They may be constructed of reinforced concrete but are usually of lattice construction. They require relatively little maintenance but, for heights above 30 m, are more expensive than stayed masts. However, lattice and concrete towers over 250 m in height are practicable and in use at a number of sites, where site size and public viewing facilities were important considerations.

Stayed masts (alternatively referred to as guyed towers) occupy a considerable land area, as the outer stay radius is between 50% and 100% of the mast height. They may be tubular or lattice. When compared with self-supporting towers, stayed masts become increasingly attractive as the height increases. A number of tubular steel masts up to 2.75 m in diameter with glass-reinforced plastic (GRP) shrouded stayed lattice extensions carrying VHF or UHF antennas with heights of 380 m are in service. However, whilst, like large concrete towers, they can accommodate lifts for access to antennas and aircraft warning lights, they have very little else to offer. They are difficult to erect, expensive, liable to vibrate, sometimes dangerous in steady wind conditions and are very much less flexible than lattice structures, to which additional antennas can be attached with ease at any height. A self-supporting tower with a height of 100 m costs around twice as much as a stayed mast of the same height.

Examples of self-supporting towers and stayed masts are shown in Figure 8.2 and most support structures for broadcast antennas are of the 'lattice' type shown in Figures 8.2(a) and 8.2(b). There is a choice of angle and tube for the cross-bracing and leg members and these may be of either steel or aluminium. Aluminium is suitable for lightly loaded masts or towers with heights up to around 70 m but for greater heights and heavy load conditions steel members are more economical and less prone to fatigue failure. This is because, to provide the required structural strength, the cross-section of aluminium has to be increased, with a resulting increase in wind load.

Aluminium members have the advantage that they do not require treatment after fabrication to inhibit or minimize corrosion. Steel members must be galvanized (i.e. coated with zinc) by the 'hot-dip' process and, if erected in a corrosive atmosphere (near the sea or in an industrialized area), they must be painted at regular intervals. If steel tube members are used they must be galvanized both internally and externally and drain holes provided to prevent the accumulation of water.

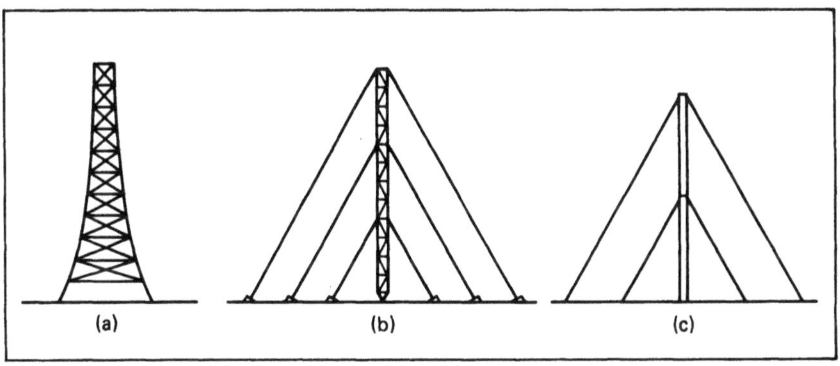

Figure 8.2 Masts and towers. (a) Lattice self-supporting tower. (b) Lattice stayed mast (guyed tower). (c) Tubular lightweight stayed mast

Structural glass reinforced plastic (GRP) cantilevers

Structural GRP cantilevers up to 15 m tall and 1.6 m diameter have been used to accommodate UHF antennas, and whilst some have been in service for about 20 years, their safe working life cannot be predicted or reviewed with the same confidence as lattice steel cantilevers.

Existing structures

Existing structures, for example water towers, chimneys and tall buildings, can sometimes be utilized, particularly if the antenna is relatively small. But when attempts to do so are made there are frequently more problems than initially envisaged, and unless there are good local facilities and structural expertise it is probably preferable for the antenna and support structure to be supplied as a package.

Document RS.226 issued by the Electronics Industries Association (EIA) in the USA provides guidance on the structural design of antenna support structures. In calculating the mechanical loading to which a support structure is subjected, the 'dead load', the 'ice load', the 'wind load' and the 'design load' must be taken into account:

1. *Dead load* – this includes the weight of the support structure, including antennas, stays and permanent maintenance facilities.
2. *Ice load* – this is the weight of ice accumulating on the support structure and antennas. The load is calculated as 897 kilograms per cubic metre of ice.
3. *Wind load* – this includes all forces generated on the structure and antennas by a specified horizontal wind pressure. When calculating the wind load, the ice coating must be included as part of the projected area.
4. *Design load* – this is the specified combination of wind, ice and dead loadings that the structure must withstand.

When calculating the effect of wind the pressure in kilograms per square metre is taken to be $P = 0.098V^2$, where V is the wind velocity in metres per second. For example, a 100 mile/hour (44 m/s) wind would result in a pressure of 190 kg/m^2. Wind pressure acts on the projected area (rather than the actual area) of a structure and the effect is therefore different for flat and cylindrical members. The pressure is taken to be $1.5P$ for open-faced lattice structures and $2/3P$ for cylindrical members. Figure 8.3 illustrates the calculation of wind pressure for a number of typical cases and it should be remembered that the wind force is applied to the whole projected area of the structure, including the antennas and ice (where applicable).

The wind velocity varies with geographical area and increases with height above ground. For the latter reason, masts and towers over 198 m in height are usually designed for a uniform wind pressure 1.5 times that used for heights below 198 m. It is important to note, however, that local structural design codes and conditions vary with the site area concerned and must be observed with a consequent effect on price.

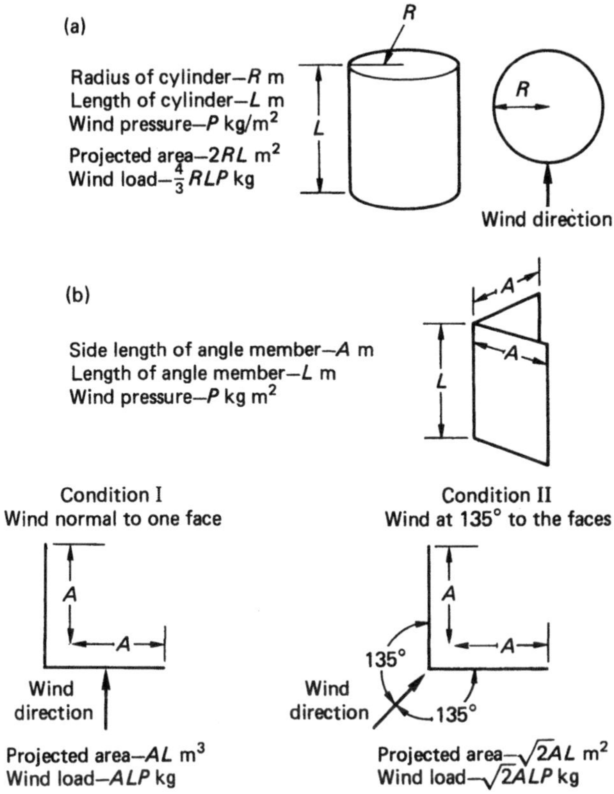

Figure 8.3 Calculation of wind pressure. (a) Cylindrical structural member. (b) Angle structural member

Any mast or tower is designed with a factor of safety so that the structure will withstand a load significantly greater than the design load. Clearly, it would be unacceptable to use a structure that was on the verge of collapse when subjected to the maximum wind and ice load expected to occur. Factors of safety (FOS) vary for different parts of the structure and typical values are as follows. These FOS figures are the amounts by which the maximum design load must be exceeded for damage to occur: mast structure, 1.6 times; stays, 2.5 times; stayblocks, 2 times (before stayblocks are lifted).

8.3.2 Stays

When an MF mast radiator is used the stays are located in the radiated RF field and, if a conducting stay material is used, currents will be induced and distortion of the radiation pattern will occur. The distortion can be eradicated by breaking up each stay into short sections (around 0.1 wavelengths at the operating frequency). Alternatively, non-conducting stays can be used but, for most applications, available materials have serious disadvantages:

1. *Polyester rope.* This material is not truly elastic in that it will continue to stretch throughout its useful life. Further, at sites where deposits of salt or dust occur the stays can become conducting when wet and this can lead to overheating and burning in the presence of strong RF fields. However, polyester stays are suitable for low-power installations using light portable masts where the stretch is relatively unimportant and the ease with which they can be terminated and joined is an advantage.
2. *Epoxy bonded glass fibre rope and rod.* These materials do not suffer from the stretch and permanent deformation occurring with polyester rope and will carry greater mechanical loads. However, they cannot be bent into a small radius and coiled for easy transport. Further, they are difficult to terminate and the epoxy bond deteriorates in the presence of salt deposits, ultraviolet light and adverse weather conditions. Also, the rope or rod may be fractured during or prior to erection. Fractures can occur that are not readily visible and may result in failure with consequent damage to the structure.
3. *Other non-conducting ropes and rods.* During recent years, other non-conducting materials have become available but none of these have proved entirely satisfactory as a stay material.

In view of the disadvantages of non-conducting stay materials for permanent installations it is necessary to use conducting materials, broken up with insulators when mast radiators are used. Probably the best solution is to be used Alumoweld (aluminium-clad steel-wire rope) which, although relatively expensive, has important advantages. Aluminium in pure or certain alloy forms is resistant to corrosion but (due to its low tensile strength) is not suitable as a stay material. Copper is also corrosion resistant but movement due to varying wind conditions causes 'work hardening' with a consequent risk of failure. Also, copper rod of a diameter providing adequate tensile strength is relatively heavy and the resulting increase in dead load means that increased strength is needed in the mast itself. There is, in fact, little doubt that the ideal stay material is Alumoweld, which has the following advantages:

1. It is eight times stronger than a pure aluminium stay of the same radius.
2. It weighs only $2\frac{1}{2}$ times as much as an aluminium stay of the same radius.
3. The corrosion resistance is extremely good and virtually the same as that of pure aluminium. It is interesting to note that any mechanical damage to the aluminium coating is 'self-healed' by spreading of the aluminium cladding.

8.3.3 Insulators for mast radiators

For light stayed masts with heights up to 70 m, simple 'rod' or 'egg' stay insulators are normally used. The former are to be preferred and the most effective types comprise a ceramic insulating rod with high-tensile aluminium fittings swaged on to each end. For higher masts and heavier loads 'fail-safe' stay insulators are necessary. These insulators are designed so that failure of the insulating material does not result in mechanical failure with consequent damage to, or collapse of, the structure. Fail-safe insulators are normally of the type shown in Figure 8.4, in which the insulating material is a ceramic tube compressed between two endpieces. The compression is provided by tension in an internal insulating strap joining the endpieces which, if the insulating tube breaks, provides a mechanical linkage. The insulating tube is filled with non-conducting oil to prevent the ingress of moisture.

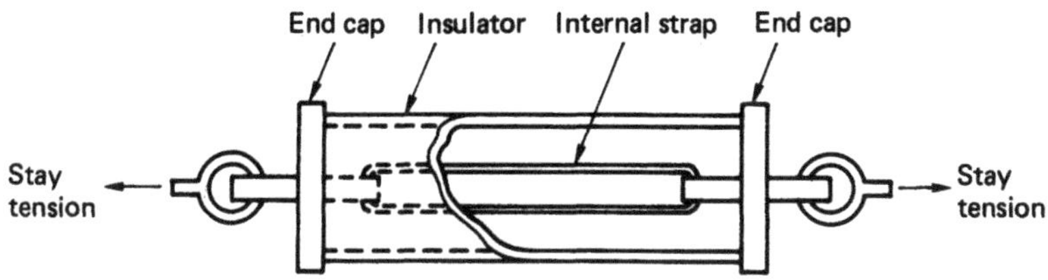

Figure 8.4 Fail-safe insulator

MF mast or tower RF radiators (unless of the shunt-fed type) must be insulated from earth. An insulator is therefore placed at the base of a stayed mast (or one insulator at the base of each leg of a self-supporting tower). These insulators must, of course, be able to carry a mechanical load (the weight of the mast or tower) in addition to their electrical function and pose a significant design problem. If mechanical strength is achieved by using a solid insulating cylinder, internal heating will occur due to dielectric losses and cracking may result. It is usual, therefore, to use a cylinder of ceramic material which is capable of withstanding the compression forces and which is fitted between metal endpieces. As in the case of cylindrical stay insulators the cylinder is filled with non-conducting oil to prevent the ingress of moisture.

8.3.4 Lightning strikes and earthing

Virtually all masts are subject to lightning strikes, and in tropical areas the incidence is high. The effects of lightning are well known and the high levels of electrical energy that are generated can result in severe damage to a transmitting installation and in danger to staff. It is not possible, of course, to prevent the occurrence of lightning strikes but their effects can be minimized by careful design. Lightning tends to strike the high well-defined point in an area, and masts or towers (particularly those sited on hills) are therefore especially vulnerable. The discharge of electrical energy associated with a lightning strike takes the form of a high-current pulse that passes from the point of the strike (the top of the mast) to points at a lower potential (usually in the neighbourhood of the ground). However, it is important to note that the route taken by the current can be via more than one path and that it is difficult to predict these paths precisely. Ideally the current should pass down mast and directly to earth but, in practice, components of current may flow to earth via the mast lighting transformer, the antenna matching circuits or the transmitter output circuit. To ensure that lightning currents flow directly to earth (so that protection is provided for both equipment and staff) it is essential that a transmitting station has a comprehensive earthing system. This system must include the 'earthy' conductors of:

1. The transmitter and associated equipment racks
2. The power supply switchboard and standby diesel generator
3. Internal and external RF feeders
4. The mast base, stays (guys) and antenna matching circuits
5. The earth system of the transmitter building.

An effective arrangement is to bond the earths of all these systems to an earth 'busbar' comprising a copper strap, three to four inches wide and braced at all joints. In turn, the busbar is bonded to a 'main station earth' which can take the form of a pit, dug, preferably, in the wettest part of the site. Where possible, the pit should be located between the mast and the building and should be between 10 and 20 feet deep. Copper rods, about eight feet long, should be driven into the floor of the pit and bonded together with copper straps which, in turn, are bonded to the earth busbar. To retain moisture, the floor of the pit should be covered by a six-inch layer of salt, which should be covered by a six-inch layer of charcoal to protect the salt against effects of heavy rain. The pit should then be filled with earth. Where the ground is rocky, it may not be possible to excavate a pit of sufficient depth and in this case several smaller pits, joined by copper straps, should be used.

Quite apart from lightning strikes, a high level of electrical activity (particularly in tropical areas) can result in the build-up of high electrostatic charges when a base-insulated mast is used. It is therefore essential that a static drain choke is connected across the mast base insulator and that it should be checked for electrical conductivity at regular intervals. The choke allows electrostatic charges to drain away and should have a DC resistance of at least 1 kΩ.

It is, of course, impossible to provide a low-impedance metallic path across the stay and base insulators of an MF mast radiator but a means of passing the current pulses due to lightning strikes must be provided. The base insulator, in particular, is subjected to very high voltages under lightning strike conditions and these can ionize the surrounding air and give rise to an arc (or flashover). A flashover may result in damage to the insulator by tracking and, in the case of high-power installations, an arc may be set up and maintained by the RF energy which can result in serious damage. For both base and stay insulators the voltage at which flashover occurs can be increased very considerably by fitting corona rings (large metallic discs with a smoothly rounded outer edge). In the case of base insulators it is also necessary to provide a 'lampshade-shaped' non-conducting rainshield to prevent rainwater flowing over the insulator surface with a consequent reduction in flashover voltage. As a final precaution, a ball-gap (two conducting spheres connected to the top and bottom of the base insulator) is provided. The balls are spaced so that the lightning voltages will set up an arc (thus protecting the insulator) but the RF voltage will not be sufficient to maintain it. Figure 8.5 shows the arrangement of the static drain choke and ball-gap for a base-insulated MF mast radiator.

The complication introduced by a base insulator can be avoided by the use of a 'shunt-feed' in which the mast base is earthed and the feeder is connected to the mast structure by means of a 'delta-feed' as illustrated in Figure 8.6. It can be shown that there is a point on the mast at which the impedance can be adjusted to match the feeder. The shunt-feed arrangement, although apparently simple, is complex in practice and has the disadvantage that radiation from the feed arrangement can distort the horizontal radiation of the antenna. Further, it is not easy to arrange for the required currents in the parasitic mast radiators for a directional MF antenna. The use of shunt-feed is mainly restricted to single-mast omnidirectional antennas installed in tropical regions subject to a high incidence of lightning strikes.

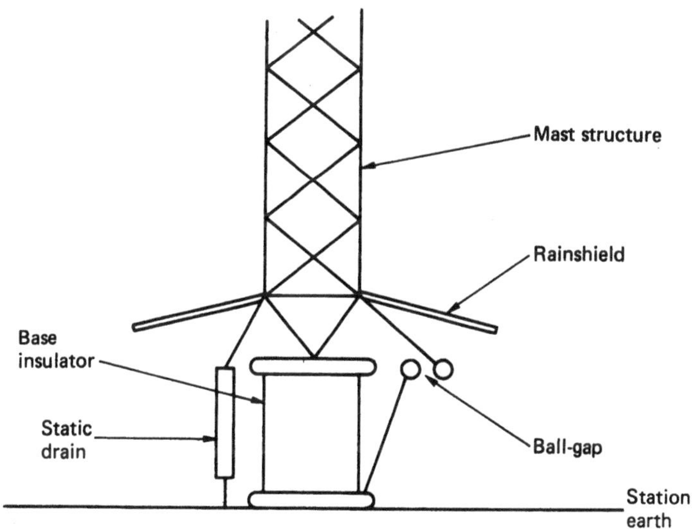

Figure 8.5 Base-insulated MF mast radiator arrangement

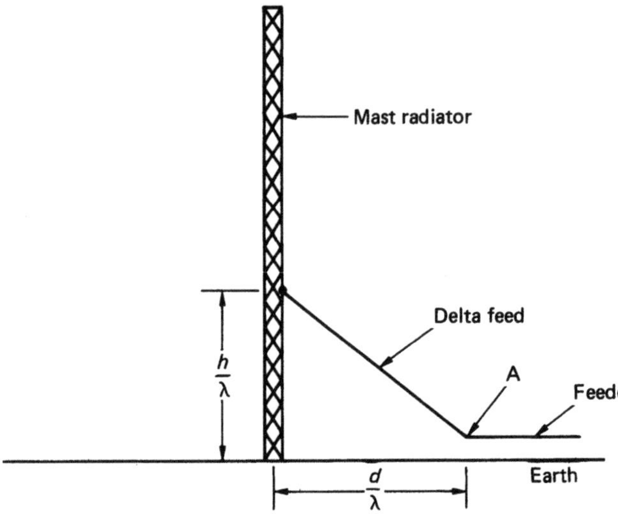

Figure 8.6 Shunt-fed grounded mast radiator. Dimensions d/λ and h/λ determine the impedance at point A

8.3.5 HF antennas

HF antenna systems of the log-periodic (LPA) and curtain types comprise support structures together with a complex arrangement of radiating conductors and support wires and present a difficult structural design problem. To achieve a design that is both safe and economic can only be achieved by computer methods as explained in Chapter 9. However, a number of points are worth noting:

1. Wires used for radiating elements and support rigging should have minimum wind area and good resistance to atmospheric corrosion. As explained in Section 8.3.2, a very suitable material is Alumoweld, except in cases where pure aluminium can provide the large radius necessary for high operating voltages.
2. Wire terminations and fittings should be of high-purity aluminium.
3. Radiating wires should be terminated in low-loss ceramic rod insulators or fail-safe types where necessary. Nuts and bolts, where used, should be of high-strength aluminium.

HF curtain antenna systems can be designed most economically from the structural point of view when a 'balance weight' system is used. The arrangement is illustrated in Figure 8.7 and it will be seen that the antenna curtain is suspended from a horizontal catenary wire which runs over pulleys attached to the masts. Balance weights are attached to each end of the catenary

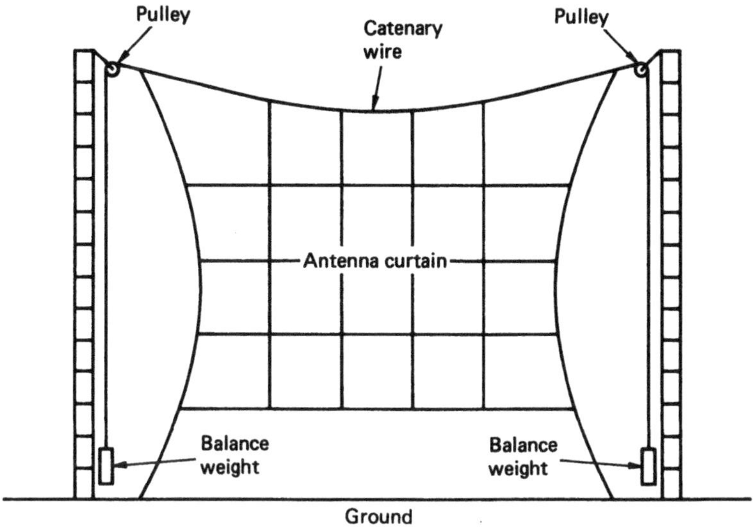

Figure 8.7 Balance-weight system for HF curtain antenna

wire to provide the correct tension in the curtain under still-air conditions. Under wind and ice conditions the balance weights lift and allow the curtain to sag so that the tension in the catenary is maintained at a constant value under all conditions and the tensions in the antenna wires are reduced. Hydraulic rams are sometimes used in place of balance weights but have the disadvantage that the available travel is limited and damage may occur to the curtain and/or catenary wire under severe weather conditions.

8.3.6 Reliability

The reliability of an item of equipment is often stated in terms of the MTBF (mean time between failures), which is based on operational experiences. For example, if one item of equipment suffers five failures in 10 years (or, alternatively, 10 similar items suffer 20 failures in four years) it can be said that the MTBF is two years.

The MTBF concept can only be applied to new items of equipment under restricted conditions. If, for example, a new design comprises a number of components and sub-units whose individual MTBFs are known, the predicted MTBF will be given by:

$$1/M = 1/M_1 + 1/M_2 + 1/M_3 \ldots 1/M_N$$

where M is the MTBF of the equipment and $M_1, M_2, M_3 \ldots M_N$ are the known MTBFs of the individual components and sub-units. This approach is not readily applicable to antennas and structures since they are not readily divisible into sub-units with known MTBFs. One way of assessing the reliability of antennas and structures offered by a single contractor is to obtain a statement of the number of antennas and structures (of all types) that are in service and the total number of faults that have occurred. The MTBF of a particular type of antenna can then be assessed as the total number of service years divided by the total number of faults and a reasonable specification would be for an MTBF of 50 years. For example, if 30 antennas of one type have been in service for four years and 50 antennas of another type for two years and three failures had occurred, the MTBF would be:

$$\frac{(30 \times 4) + (50 \times 2)}{3} = 73 \text{ years}$$

Chapter 9

Computer methods

9.1 Introduction

The electrical performance of an antenna is governed by Maxwell's equations, which control the relationship between the input power and the radiated power but, in practice, it is difficult to obtain a direct solution in any but the most simple cases. If the current distribution on the conductors of an antenna is known it is relatively easy to determine the nature of the radiated field. The problem, however, is to find the current distribution and this appears in the form of an equation that cannot, in most cases, be solved by normal mathematical methods. As a result, until relatively recently, antenna design was based on approximate solutions of Maxwell's equations backed up, where necessary, by tests on a scale model of the antenna or tests on the modification to the antenna after installation at site. These procedures are usually laborious, expensive and time-consuming.

As an example of the antenna design problem it is useful to consider the relatively simple case of the monopole radiator shown in Figure 9.1(a) (in practice this could be an MF mast radiator). Maxwell's equations show that the distribution of current on the monopole is approximately sinusoidal (zero at the top of the monopole and reaching a maximum at a distance of a quarter of a wavelength from the top with another zero a half-wavelength from the top). Using the sinusoidal current approximation the vertical radiation pattern (VRP) of the antenna can be calculated and will be as shown in Figure 9.1(b). However, using the sinusoidal current approximation, the input impedance will be a pure reactance, which is obviously incorrect as, if it were true, the antenna would radiate no power. It will also be seen that there is a zero in the VRP shown in Figure 9.1(b) which is of importance in the design of MF anti-fading antennas. This is because the zero can be adjusted so that the sky-wave is eliminated at the edge of the service area and the coverage is thereby increased. In practice, however, the radiated field strength does not fall to zero as is indicated by the sinusoidal current distribution and this affects the anti-fading performance of the antenna. It is therefore clear that, even in the case of a simple monopole antenna, the predictions of the VRP and input impedance based on the sinusoidal current approximation are not sufficiently accurate and a more rigorous solution of Maxwell's equations is necessary if design errors are to be avoided. An analogous problem arises in the mechanical and structural design of antennas, particularly those operating in the HF band. The equations governing the stresses in the masts, stays, radiator wires and support wires are such that an

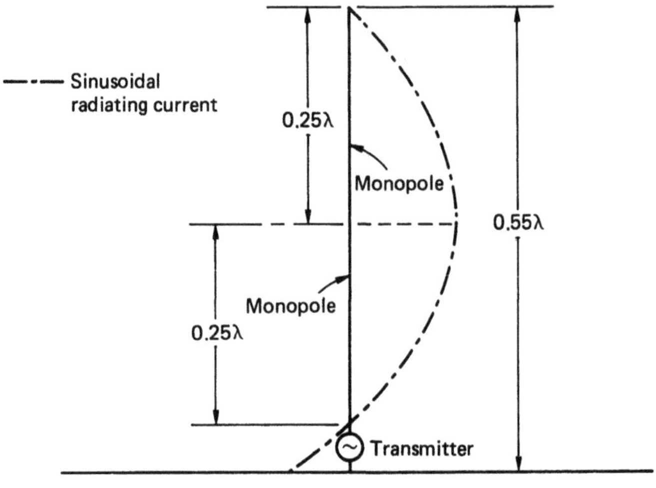

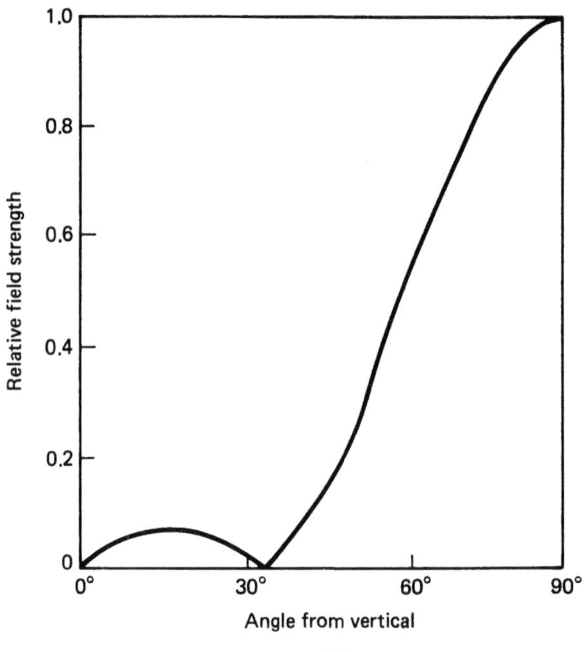

Figure 9.1 Monopole with sinusoidal current distribution. (a) Current distribution. (b) Vertical radiation pattern

accurate solution is impractical under wind conditions and difficult even under still-air conditions.

The advent of the high-speed computer has made it possible to obtain accurate solutions to the problems encountered in both the electrical and mechanical and structural design of antennas. There are a number of computer programs for the electrical and structural design of antennas and, in this chapter, the programs devised by Technology for Communications International (TCI) are used as a basis of discussion. These programs are WIRA (for electrical design) and STRUC (for structural design) and are particularly flexible in operation, enabling a wide range of problems to be solved rapidly, economically and accurately.

9.2 Electrical design of antennas

9.2.1 Antenna electrical design programs

Antennas for use in the LF, MF and HF bands comprise, essentially, arrangements of cylindrical radiating elements. In the HF band wire conductors are always used but in the LF and MF bands the radiating conductor usually takes the form of a mast or tower structure. Like the wires used in the LF and MF bands, cross-sectional dimensions of masts and towers in the LF and MF bands are small compared with a wavelength at the operating frequency and, without significant loss of calculation accuracy, they can be considered as cylinders. Thus, for the purposes of analysis, LF, MF and HF antennas can be considered as arrangements of interconnected cylindrical conductors of varying lengths and diameters.

The electromagnetic field radiated by an arrangement of wire conductors can be calculated without difficulty when the current distribution is known. For the purposes of calculation, the wires can be divided up into a large number of small elements, each of which acts as an elementary antenna, and the total radiated field determined by adding the radiating fields of all of these elements. However, an accurate determination of the current distribution presents a much more difficult problem, since all that is known is that this distribution must meet two ‘boundary conditions’:

1. The component of the electric field generated by the currents in all the radiating elements and lying along the surface of any conductor (the tangential field) must be zero. Put in, perhaps, more familiar terms, one cannot have a voltage across a short-circuit.
2. The current distribution must take account of voltage sources (i.e. the transmitter) and lumped electrical components (e.g. insulators) connected to the radiating elements.

These boundary conditions can be stated in the form of an ‘integral equation’, the solution of which gives the current distribution at all points on the radiating elements. The integral equation, despite its rather formidable name, is simply a statement that, at any point in the antenna, the tangential fields due to all the current elements must add up to zero at each point on each conductor. Alternatively, at the terminals of each lumped impedance or generator the values of current and voltage must correspond to the impedance or generator voltage.

The solution of the integral equation gives the current distribution (the current at each point on each conductor) and, in addition, enables the input impedance, radiation pattern and other important properties of a wire antenna of arbitrary geometry to be determined. Here it is worth pointing out that the accurate determination of input impedance is the most stringent test of a method of calculation. The monopole example of Figure 9.1, where the sinusoidal approximation to the current distribution gives a purely reactive (and incorrect) input impedance, is an example of this. For example, using the sinusoidal current approximation for a quarter-wave monopole will give the radiation pattern with reasonable accuracy but will predict zero input impedance (instead of the correct value of resistance in the region of 37 Ω). Basically, the WIRA computer program solves the integral equation, and determines the current distribution by dividing each wire of the antenna into a number of intervals. As a result the integral equation (which, as has been pointed out, is nearly always insoluble) is replaced by a series of linear simultaneous equations for the values of current at the centre of each interval. The solution of these simultaneous equations would involve a prohibitive amount of work using a slide rule or handheld calculator but, with a high-speed computer, can be done in minutes even for complex antennas. Clearly, the accuracy of the method depends on the number of intervals into which the wires are divided, and the greater the number of intervals, the greater will be the number of simultaneous equations and the longer will be the calculation time. However, the number of intervals into which each conductor is divided need not be unduly large. For example, a quarter-wave monopole above a conducting plane can be represented by three sections each with a current distribution arranged to avoid discontinuities at the junctions between them. All conductors comprising a practical antenna (including radiating elements, support wires, interconnecting feeders and support structures) can be included in the program together with the effects of lumped components (including insulators). It must be stressed that the numerical solution of the integral equation is, in the limit, an exact method and that the principal errors that arise in the calculation of antenna performance come from approximations in the integration process. By increasing the number of intervals into which the conductors are divided these errors can be reduced to any level that is economically acceptable. In practice, it is quite easy to obtain accuracies better than those obtained by actual measurement and at a much lower cost.

9.2.2 The design of LF, MF and HF antennas

In many applications in the LF, MF and HF bands antennas are located above ground of finite conductivity which absorbs part of the power radiated by the antenna and distorts the radiation pattern. Electrical design programs can take account of finite conductivity in the region of the antenna and determine, accurately, the radiation efficiency and the radiation pattern shape. As the input generator voltage is input to the program, and the current at the input terminals is then determined, the impedance at the driving-point (as a series connected resistance and reactance) is given directly. As has been made clear, because the method of analysis is fundamentally accurate the input impedance is given with a degree of accuracy dependent on the number of intervals into which the conductors are divided.

These programs are arranged so that it is possible to optimize the design in favour of a particular parameter. For example, it is possible to optimize a design for maximum gain, maximum bandwidth or maximum front-to-back ratio. Important parameters calculated by the WIRA program are the radial electric field (as opposed to the tangential field already discussed) at the surface of each wire, the voltage across each insulator and the maximum current in each wire. These parameters indicate whether flashover or overheating is likely to occur and are therefore of great importance for high-power applications.

To demonstrate that the computer results are reliable it is of interest to compare the calculated radiated pattern and impedance of antennas that can be scaled accurately and, therefore, measured accurately. As has been pointed out, a comparison of calculated and measured impedance is the most sensitive test of the computer program, as it depends more strongly on approximations than do the current distribution and radiation pattern. Figures 9.2 and 9.3 show a comparison between the calculated and measured impedances of a scale model

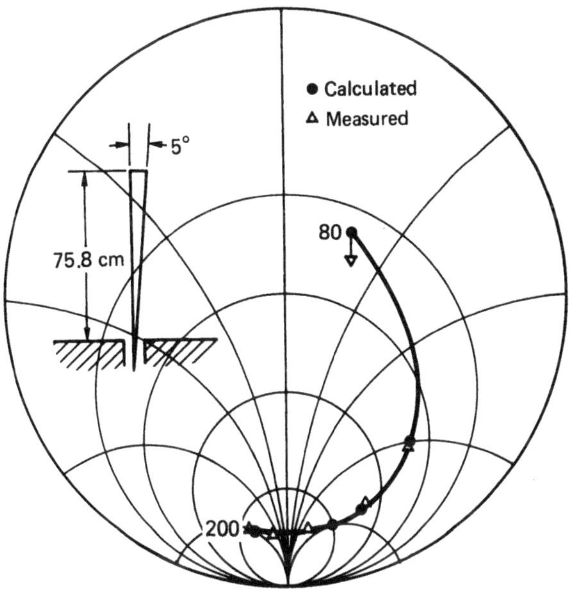

Figure 9.2 Tapered monopole

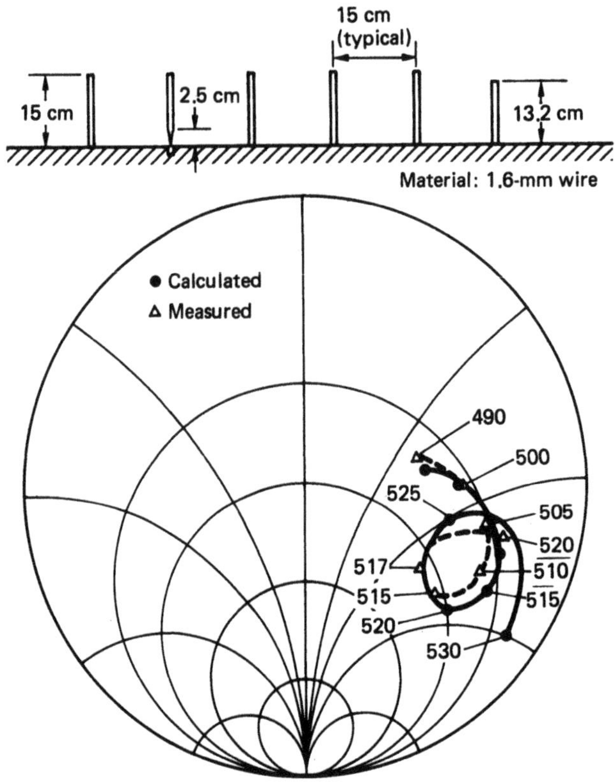

Figure 9.3 Yagi antenna

tapered monopole and six-element Yagi antenna. In both cases, the agreement is seen to be good. Indeed, the difference between the calculated and measured values could well be due to measurement errors rather than errors in the calculation process.

9.2.3 The design of VHF, UHF and SHF antennas

Early electrical design programs were originally developed to solve problems arising in the design of wire antennas (i.e. antennas for the LF, MF and HF bands that comprise conductors in the form of cylindrical wires and conductors). However, the principle of dividing a conductor into sections to permit calculation of the current distribution can be readily extended and applied to conducting surfaces. For example, the parabolic reflector of a microwave dish antenna can be divided into a number of small elements of surface and the linear simultaneous equation method then applied to obtain the current distribution.

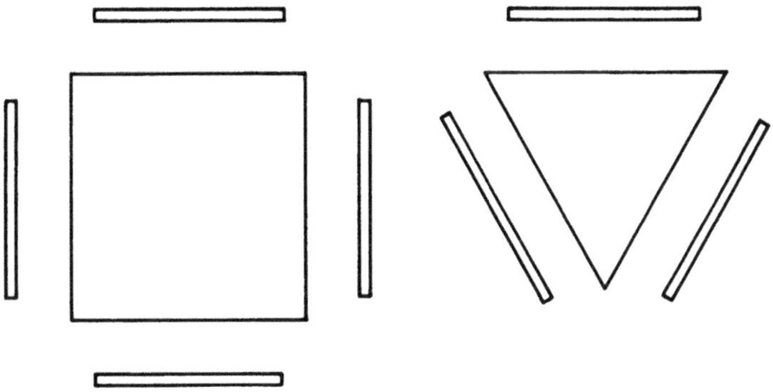

Figure 9.4 Panel antennas on square and triangular masts

Certain problems arising in the design of VHF and UHF transmitting antennas are particularly intractable to normal methods of analysis and an example of this is the case of a panel antenna (see Chapter 4) on a support mast of square or triangular cross-section as illustrated in Figure 9.4. In this case the problem is that the cross-sectional dimensions of the mast cannot be regarded as small compared with a wavelength and cannot, therefore, be replaced by a single cylindrical conductor. However, the screen of the antenna will comprise parallel wires (either vertically or horizontally disposed, depending on the polarization) and can, therefore, be solved by the WIRA wire antenna program.

As in the case of LF, MF and HF bands, antennas in the VHF, UHF and SHF bands can be modelled at a higher frequency. For example, a VHF/FM antenna operating at 100 MHz with a height of 30 m (10 wavelengths) could be modelled at 1000 MHz with a height of 3 m. However, as in the case of the lower bands, it is difficult to construct a scale model with sufficient accuracy (for example, the wires in the screen of a panel antenna may be too thin for constructural convenience). Thus a computer solution offers the advantages of speed and accuracy when an antenna of this type is to be designed.

9.3 Structural and mechanical design of antennas

9.3.1 Structural design problems

When the electrical design of an antenna has been determined, a satisfactory mechanical and structural design must be evolved. Particularly in the case of HF high-gain antenna arrays, the mechanical and structural design presents a problem comparable in complexity to that posed by the electrical design. The problem is, essentially, one of determining a structural arrangement which satisfies the electrical requirements and, at the same time, provides adequate, but economic, structural safety. Prior to the advent of the high-speed digital computer it was quite possible to design a safe structural arrangement for any antenna. However, since accurate calculation was not possible, the designer had to play for safety and, as a result, the structure was 'over-designed'. In other words, the design was usually stronger than it needed to be and, as a result, was more expensive than it needed to be.

The wind pressure acting on a structure can be represented as a number of horizontal forces (their magnitudes depending on wind speed as a function of height), as illustrated in Figure 9.5(a) for an MF mast radiator, and it is then possible to calculate the stresses in the mast structure and the stays. Such a simple method of analysis does not take into account the bending of the structure under high-wind conditions and this results in significant changes in the stresses, as illustrated in Figure 9.5(b). It is clear, therefore, that an accurate analysis must take account of mast and antenna deformation under wind conditions and this is only possible with a sophisticated structural computer program.

9.3.2 Structural design programs

An accurate structural design process for a mast–antenna system must take account of all displacements, forces and moments in an arbitrary arrangement of interconnected wires, cables and structural members under arbitrary combinations of gravity, wind loading and ice loading. In most modern programs the input information, briefly, is as follows:

1. A specification of the junction points at which the members and/or wires are joined (these are known as 'node locations').

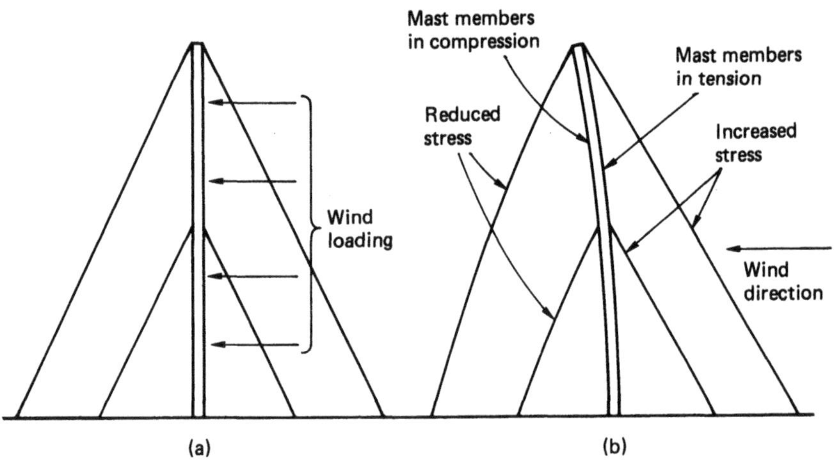

Figure 9.5 Effect of wind on mast structures. (a) Simple analysis. (b) Accurate analysis

2. A specification of the constraints to be imposed on node movements, i.e. the limits on the movement of junction points that is to be permitted under wind and ice loading conditions.
3. Structural descriptions of the wires and structural mast members. These describe the way in which they become deformed under compression, tension, etc.

The program is normally operated in two stages:

1. The stresses on the structure are determined under still-air, dead-load conditions and design modifications made until a satisfactory configuration has been achieved (satisfactory from both the electrical and structural points of view).
2. The lengths of wires and members that are under tension under still-air, dead-load conditions are set to take account of the tension and various combinations of wind loading and ice loading are applied. Design modifications are then made, as necessary, to eradicate structural weaknesses that develop under dynamic load conditions.

(a) Specification of the computer program

In the design process, the nodes (junctions) may be forbidden to move or be permitted to move along a line, in a plane, or be completely free. In describing antenna and structural components, properties are specified that include diameter, weight per unit length, elastic stiffness, temperature coefficient of expansion and (for rigid members) bending stiffness and torsional rigidity. Further, it is possible to specify that, instead of the length, the tension in a member is held constant during calculation of the displacements. Rigid members may be considered as rigidly attached at their ends or, alternatively, by means that permit rotation about the node points (i.e. with 'pin joints'). In addition to the antenna system configuration there are environmental data which include ice loading, wind velocity and direction (both in azimuth and height).

(b) Program operation

The wind force on each member is calculated from the properties of standard air as modified by any density factor and standard published data on lift and drag coefficients for cylinders (for wires and cables) and flat plates (for rigid members). The force is calculated at the mid-point of each member and then distributed over its length. Ice is added to the cables, wires and rigid members to the thickness specified, thus automatically increasing the diameter and the weight.

The program calculates the force–moment–displacement–rotation relationship in the neighbourhood of the current positions of the nodes and then, simultaneously, displaces and rotates all movable nodes in such directions as to tend towards equilibrium. The cycle of calculations is then repeated using the new node positions. This process (which is essentially an 'iterative' process as described in Section 9.4) is continued until all unbalanced node forces are reduced below a specified magnitude.

(c) Results and approximations

The solution obtained for an antenna system consists of the node (junction) displacements, forces, moments, tensions and sags for cables and end-point forces and moments for rigid members. The approximations used in the program are only algebraic and geometrical in nature. Second-order effects (which may become important when there are large sags in cables or near-buckling conditions in rigid members) can be eliminated by sub-dividing the members concerned into two, or more, smaller members for the purposes of calculation. As a result, the solution approached by the iterative process can be made rigorously accurate by increasing the number of sections into which members are divided and, consequently, increasing the calculation time.

Rigid members having complicated mechanical properties (e.g. with different stiffness parameters in different directions) can be simulated by means of a combination of simpler members with differing characteristics.

9.3.3 Advantages of computer design

As is illustrated by the descriptions of electrical and structural design programs, high-speed computers enable engineering problems of considerable complexity to be treated by very general theoretical methods. In addition to the advantages of speed, economy and flexibility, when compared with experimental methods, the numerical approach provides answers with precision and degree of detail beyond the reach of most experimental techniques. As a result, computer-aided antenna designs result in an antenna whose performance is predicted with a high degree of accuracy and which can be installed without the need for any modification after installation at site. Further, the accuracy of the mechanical and structural design is such that 'over-design' is avoided and the antenna system is the most economical that can be achieved.

9.4 Computer solutions using a PC

Such computer programs as those described in Sections 9.2 and 9.3 involve a large number of parameters and a very large number of calculations and can only be handled by a mainframe computer. However, there any many propagation and antenna design problems that can be handled by a conventional

PC (personal computer). Many calculations that, a few years ago, required days or weeks of labour with a handheld or mechanical calculator can be carried out in a few minutes. It is, of course, necessary to program the PC for any particular problem but, for a calculation that needs to be carried out frequently (e.g. MF or HF propagation analysis), this is well worthwhile in terms of overall time saved.

9.4.1 General solutions

There is a wide range of software (programs) available for PCs but this is mainly for business and administrative purposes. Problems in the antenna and propagation field are of a specialized nature and it is not often that standard software is available for the range of problems that generally occur. However, a number of similar mathematical formulae and operations arise in a wide range of problems and general solutions can be programmed for inclusion in programs aimed at solving specific problems. This obviously simplifies the writing of programs for the solution of a wide range of problems.

The range of general solutions useful for inclusion in antenna design and propagation software is very wide but two examples should serve to illustrate the method.

(a) Integration

Many problems, particularly in the field of antenna design, present themselves in the form of an integral between fixed limits. For example, the gain of an antenna can be determined by integrating the radiated power over the complete range of azimuth and elevation angles. For solution by computer the range of integration is split up into a number of small intervals so that the integral becomes the sum of the areas of a number of strips as shown in Figure 9.6. The areas of strips (the elemental areas) are then added to give the approximate value of the integral. This is known as 'arithmetic integration' and the accuracy of the answer clearly depends on the number of intervals into which the range of integration is divided. The accuracy can be improved by using Simpson's rule, in which the strips are taken in pairs so that three ordinates are used in the calculation of each elemental area. Simpson's rule can be used to reduce the number of strips required for a given level of accuracy or to reduce the number of strips and hence reduce the time required for the calculation.

Figure 9.7 lists an integration program, using Simpson's rule, for the integration of $1/x$ (the solution is, of course, known to be ln (x) where ln is the natural of naperian logarithm). Over the integration interval 1 to 3, and using 10 intervals (line 30), the answer is given accurately to five decimal places in less than one second. The program is general in the sense that the function to be integrated is contained in a subroutine and can be changed by changing line 1011. For example, if the function required were to be tan X/X then line 1011

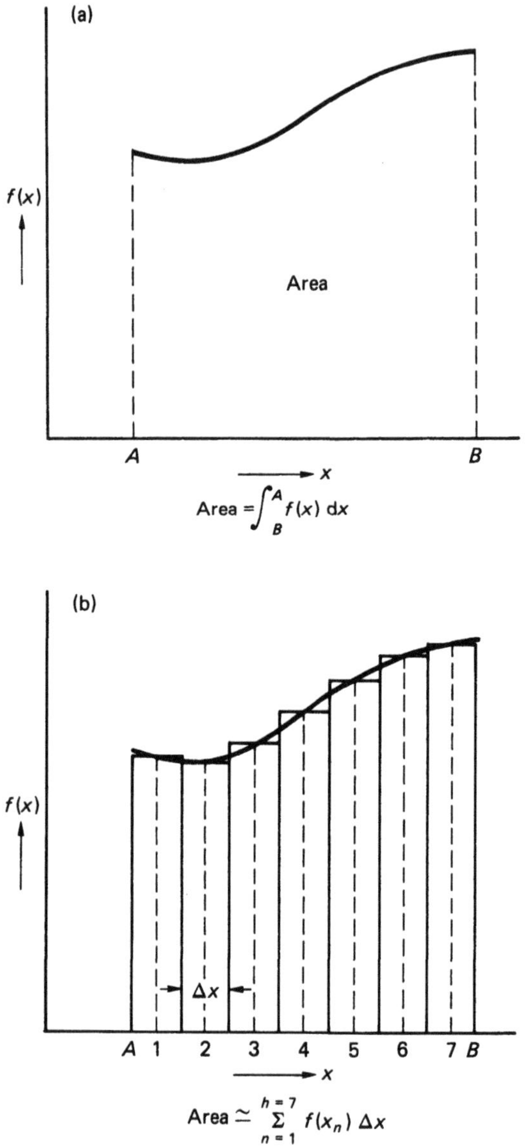

Figure 9.6 Arithmetic integration

would be changed to $y = \tan X/X$. These examples are simple functions but line 1011 could be replaced by a much more complicated function and, for example, calculate the mutual impedance between a pair of antennas or the attenuation of a signal by diffraction over a knife-edge.

The program in Figure 9.7 is in BBC Basic but can be readily modified for other versions of Basic or other commonly used languages.

```
>LIST
  10 INPUT ''LOWER LIMIT OF VARIABLE '';ll
  20 INPUT ''UPPER LIMIT OF VARIABLE '';ul
  30 INPUT ''NUMBER OF INTERVALS'';int
  40 LET h=(ul-ll)/int
  50 LET m=0
  60 LET area=0
  70 REM CALC OF AREA
  80 REPEAT
  90 LET x=ll+m*h
 100 GOSUB1000
 110 LET ordi=y
 120 LET x=ll+m*h+h/2
 130 GOSUB1000
 140 LET ord2=y
 150 LET x=ll+m*h+h
 160 GOSUB1000
 170 LET ord3=y
 180 LET delta=(ord1+4*ord2+ord3)/6*h
 190 LET area=area+delta
 200 LET m=m+1
 210 UNTIL m>int-1+0.1
 220 PRINT
 230 PRINT area
 240 END
1000 REM SBR OF FUNCTION
1011 LET y=1/x
1020 RETURN
```

Figure 9.7 Integration program using Simpson's rule

(b) Iterative calculations

A wide range of formulae that give the answer as a function of a variable can readily be programmed for a PC. However, in many cases, it would be more useful to start with the answer and find the corresponding value of the variable. This is often difficult as it is impossible to 'invert' the formula so that the variable becomes the answer and the answer becomes the variable. For example, CCIR formulae give the value of ground-wave field strength at a distance from an MF transmitter whereas it is often required to find the distance at which there is a given field strength. This can be readily handled by a PC using a repeated series of calculations (known mathematically as an 'iterative process') which, for the case of ground-wave field strength, can be summarized as follows:

1. The distance is increased from zero in 10-km steps until the field strength is below the required value of field strength. Suppose that the distance at which this occurs is D km. D will, of course, be a multiple of 10.
2. The distance is then increased from D to 10 km in 1-km steps until, once again, the field strength is below the required value of field strength. This obviously gives the distance at which the required field strength occurs to within an accuracy of 1 km. Suppose that the distance at which this occurs is D_1. This will be a whole number of kilometres.
3. Step (2) can obviously be repeated by increasing the distance from D_1 to 1 in increments of 0.1 km, which will give the answer within an accuracy of 0.1 km. However, an accuracy to within 1 km is usually acceptable as it is within the accuracy with which the formula predicts the field strength.

```
MF GROUNDWAVE FIELDSTRENGTH CONTOURS AND FADING ZONE

FREQUENCY IN KHZ:                    800.00
TRANSMITTER POWER IN KW:             50.00
ANTENNA GAIN IN DBM:                 0.00
GROUND CONDUCTIVITY IN MS/M:         10.00
GROUND PERMITTIVITY:                 4.00
ANTENNA HEIGHT IN WAVELENGTHS:       0.25
GEOMAGNETIC LATITUDE IN DEGREES:     45.00

DISTANCE TO FIRST CLASS F/S (75.00 DB) CONTOUR: 87.00 KM

DISTANCE TO 2ND CLASS F/S (65.00 DB) CONTOUR: 140.00 KM

INNER LIMIT OF FADING ZONE: 143.000 KM

OUTER LIMIT OF FADING ZONE: 165.00 KM
```

Figure 9.8 Printout of iterative MF field strength program

The iterative process sounds complicated and, if carried out using a handheld calculator, would take a very long time to complete. However, a PC will do the calculations in a few seconds. As an example, Figure 9.8 is a printout of an iterative program that determines the distances at which field strengths of 65 and 75 dBμ occur and the distances at the inner and outer limits of the fading zone occur. The parameters used in this example are:

Operating frequency	800 kHz
Transmitter power	50 kW
Antenna height	0.25 wavelengths
Antenna gain	0 dBM
Radiation pattern	omnidirectional
Ground conductivity	10 mS/m
Latitude	45°

The calculation time for this printout using a BBC Master computer is less than 10 seconds.

9.4.2 Propagation programs

Computer programs for propagation analysis in the HF, VHF/UHF and microwave bands are described, briefly, in Chapters 3, 4 and 5 respectively. All these programs can be handled by a PC and their use results in a great improvement in the speed with which station planning can be carried out. As an example of the use of a PC for propagation analysis it is useful to consider the case of MF ground-wave propagation which can be based on a number of formulae that give the field strength as a function of transmitter power, distance and a number of path parameters. The formulae are approximate but give the field strength with an accuracy sufficient for all practical planning purposes. The field strength is calculated as the product of three factors that take account of various aspects of propagation. These factors are the attenuation that would occur over a flat earth, the additional attenuation due to the curvature and the transmitter power and antenna gain:

(a) Attenuation over a flat earth

$$x = 18s/F$$

where s is the ground conductivity in mS/m, and F is the frequency in MHz.

$$b = \tan^{-1}[(e+1)/x]$$

where e is the ground permittivity.

$$k_1 = 10.472$$
$$k_2 = k_1 dF$$
$$p = (k_1 k_2 \cos b)/x$$

where d is the distance in km.

$$k_3 = 0.3$$
$$k_4 = 0.6$$
$$A_1 = (2 + k_3 p)/(2 + p + k_4 p^2)$$
$$k_5 = (p/2)^{1/2}$$
$$k_6 = e^{-5/8} p$$
$$A_2 = k_5 k_6 \sin b$$
$$A_3 = A_1 - A_2$$

where A_3 is the flat earth attenuation factor.

(b) Earth curvature factor

$$k_7 = 0.0353$$
$$Z = k_7 d \lambda^{-1/3}$$

where λ is the wavelength in m.

$$A_4 = (1 + 2Z)^{1/2} e^{-z}$$

where A_4 is the earth curvature factor.

(c) Transmitter power

$$k^8 = 300\,000 W^{1/2}$$

where W is the transmitter power in kW.

(d) Field strength

$$\text{Field strength} = 20 \log (k_8 A_3 A_4/d) + g$$

where g is the antenna gain in dBM.

These formulae, although relatively complicated, can be easily handled by a PC using a short program. The use of field strength curves and the need to interpolate where particular frequencies and ground conductivities fall between the curves is eliminated. Programs for the formulae have been written by a number of authorities and the listing of a typical example is shown in Figure 9.9.

```
>LIST
  10 PRINT ''PROG4 - MF GROUNDWAVE PROPAGATION''
  20 @%=&20315
  70 PRINT '' ''
  80 INPUT ''FREQUENCY IN KHZ:                          '';FM
  81 LET F=FM/1000
  90 INPUT ''GROUND CONDUCTIVITY IN MS/M:               '';C
 100 INPUT ''TRANSMITTER POWER IN KW:                   '';W
 110 INPUT ''ANTENNA GAIN IN DBM:                       '';G
 120 INPUT ''GROUND PERMITTIVITY:                       '';E
 130 INPUT ''MINIMUM DISTANCE IN KM:                    '';D
 140 INPUT ''DISTANCE INCREMENT IN KM:                  '';I
 150 INPUT ''MAXIMUM DISTANCE OF INTEREST IN KM:        '';MD
 160 PRINT '' ''
 170 PRINT '' ''
 175 PRINT ''       DISTANCE KM         FIELD-STRENGTH dBU''
 176 PRINT
 180 REPEAT
 200 LET X = (18*C/F)
 210 LET B = ATN((E+1)/X)
 220 LET P=PI*D*1000*F*COS(B)/300/X
 230 LET A1 = (2+0.3*P)/(2+P+0.6*P ^2)
 240 IF P>50 LET A2=0 ELSE A2=SQR(P/2)*SIN(B)/EXP(5*P/8)
 250 LET A3 = A1 - A2
 260 LET Z = 0.0353*D/((300/F) ^0.3333)
 270 LET A4 = ((1+2*Z) ^0.5)*EXP(-Z)
 280 FIELD = 20*LOG(300000*SQR(W)*A3*A4/D) + G
 290 PRINT D,FIELD
 300 LET D = D + I
 310 UNTIL D>MD
 320 END
```

Figure 9.9 Listing of ground-wave propagation program

Figure 9.10 shows a printout of this program for a typical case of MF transmission. This is a simple example and a PC will handle much more complex calculations, including sky-wave transmission and fading zone, and Figure 9.8 is an example of a comprehensive program of this type. The program illustrated in Figure 9.9 is in BBC Basic but can be readily modified for other languages.

```
>RUN
PROG4 - MF GROUNDWAVE PROPAGATION

FREQUENCY IN KHZ:                         ?860
GROUND CONDUCTIVITY IN MS/M:              ?7
TRANSMITTER POWER IN KW:                  ?50
ANTENNA GAIN IN DBM:                      ?1
GROUND PERMITTIVITY:                      ?4
MINIMUM DISTANCE IN KM:                   ?10
DISTANCE INCREMENT IN KM:                 ?10
MAXIMUM DISTANCE OF INTEREST IN KM:       ?100

            DISTANCE KM               FIELD-STRENGTH dBu

                10.000                   105.082
                20.000                    96.406
                30.000                    90.367
                40.000                    85.600
                50.000                    81.638
                60.000                    78.243
                70.000                    75.268
                80.000                    72.617
                90.000                    70.220
               100.000                    68.029
```

Figure 9.10 Printout of ground-wave propagation program

9.4.3 Antenna design programs

As has been made clear, the accurate calculation of antenna impedance requires a sophisticated computer program. However, in the case of an omnidirectional antenna, it is possible to obtain a useful approximation value for the input impedance using a program that can be handled by a PC. An approximate estimate of the impedance is often sufficient in practice since it is relatively inexpensive to provide an impedance matching network that can be adjusted to cater for a range of impedances. Figure 3.3 shows an approximate equivalent circuit for an omnidirectional mast radiator based on a uniform transmission line (a sinusoidal current approximation) and this can be used as the basis of a program that can be readily handled by a PC. The printout of such a program is

```
ANDES1: SINGLE-MAST MF ANTENNA
        GAIN AND INPUT IMPEDANCE

RADIUS OF MAST IN M:                            0.30
HEIGHT OF MAST IN M:                            100.00

FREQUENCY IN KHZ:                          880.00
DRIVEN MAST IMPEDANCE, OHMS                91.81+j140.19
GAIN OF ANTENNA dBi/dBM:                   5.49/0.71

FREQUENCY IN KHZ:                          890.00
DRIVEN MAST IMPEDANCE, OHMS                96.11+j147.30
GAIN OF ANTENNA dBi/dBM:                   5.50/0.73

FREQUENCY IN KHZ:                          900.00
DRIVEN MAST IMPEDANCE, OHMS                100.64+j154.45
GAIN OF ANTENNA dBi/dBM:                   5.52/0.75
```

Figure 9.11 Drive-point impedance of MF mast radiator

```
ANDES7: DESIGN OF MF ANTENNA T MATCHING CIRCUIT

NOTE: PROGRAM GIVES PARAMETERS OF T-MATCHING NETWORK
      WITH SERIES INDUCTANCES AND SHUNT CAPACITANCE

CARRIER POWER IN KW ?50
ANTENNA SERIES RESISTANCE AT CARRIER IN OHMS ?19.9
ANTENNA SERIES REACTANCE AT CARRIER IN OHMS ?–77.8
ANTENNA SERIES RESISTANCE AT + MODULATION FREQUENCY IN OHMS ?20.4
ANTENNA SERIES REACTANCE AT + MODULATION FREQUENCY IN OHMS ?–75
ANTENNA SERIES RESISTANCE AT – MODULATION FREQUENCY IN OHMS ?19.5
ANTENNA SERIES REACTANCE AT – MODULATION FREQUENCY IN OHMS ?–80.7
REQUIRED INPUT RESISTANCE OF MATCHING NETWORK IN OHMS ?50
CARRIER FREQUENCY IN KHZ ?700
MODULATION FREQUENCY IN KHZ ?6

INPUT IMPEDANCE AT CARRIER:               50.00+j0.00 ohms
VSWR AT CARRIER:                          1.00:1

INPUT IMPEDANCE AT CARRIER PLUS MOD:      46.17–j8.53 ohms
VSWR AT CARRIER PLUS MOD:                 1.21:1

INPUT IMPEDANCE AT CARRIER MINUS MOD:     49.70+j9.93 ohms
VSWR AT CARRIER MINUS MOD:                1.22:1

SERIES OUTPUT INDUCTANCE                  24.86 oH
SHUNT CAPACITANCE:                        7207.93 pf
SERIES INPUT INDUCTANCE:                  7.17 uH

MAX RMS VOLTAGES AND MEAN RMS CURRENTS IN VOLTS
AND AMPS FOR PROGRAMME WITH 100% PEAK MODULATION

ANTENNA BASE CURRENT IN AMPS              65.16
VOLTS ACROSS ANTENNA BASE                 8050.63
CURRENT IN OUTPUT REACTANCE IN AMPS       65.16
VOLTS ACROSS OUTPUT REACTANCE             10961.80

CURRENT IN INPUT INDUCTANCE IN AMPS       41.11
VOLTS ACROSS INPUT INDUCTANCE             1994.99

VOLTS ACROSS SHUNT CAPACITOR              3738.98
CURRENT IN SHUNT CAPACITOR IN AMPS        77.05
```

Figure 9.12 Printout of MF antenna matching circuit program

shown in Figure 9.11 for the case of a mast radiator of 100 m height and a carrier frequency of 890 kHz. As will be seen, the program is arranged to give the drive-point impedance for a chosen range of frequencies and in Figure 9.11 these frequencies represent the carrier (890 kHz) and the extremes of the modulation band (880 and 900 kHz) assuming a maximum modulating frequency of 10 kHz.

A useful companion program to the one described for the drive-point impedance of a mast radiator is a program for the calculation of the components of an MF antenna matching network. The calculations necessary to determine the reactive impedances and their component values and the maximum voltages and currents that they must withstand can be readily included in quite a short program for a PC and an example of a printout is shown in Figure 9.12. The program is based on a 'three-element T' network and the example is based on the case shown in Figure 2.4. The program first determines the values of the three reactive impedances for the T network and then calculates the peak RMS voltages and mean RMS currents that they will need to withstand. Each inductance will be of the 'tapped' type and each capacitor will comprise a number of capacitors in parallel so that the network can be conveniently adjusted on site for optimum matching.

Chapter 10

Satellite broadcasting and high-definition television

10.1 Introduction

The number of terrestrial broadcasting stations that can be accommodated in the area of a country is limited (for example, no more than four television programmes with national coverage are possible in the UK with the allocated frequency spectrum in the UHF band). To provide the public with additional television programmes, therefore, transmissions must be made in previously unused frequency bands. The frequency bands up to, and including, the UHF band offer little scope for more transmissions and it is therefore necessary to make use of the higher frequency SHF band (3–30 GHz) to provide additional programmes. The SHF band is unsuited to other than restricted low-power terrestrial coverage (the coverage provided is too 'patchy', resulting in gaps in the service area) but is suited to DBS (direct broadcasting by satellite). Here it should be pointed out that the limitation of the number of programmes due to spectrum limitation is more severe for television broadcasting than for radio broadcasting. Primarily, therefore, broadcast transmissions in the SHF band are used for television coverage.

Strictly speaking, DBS falls outside the scope of terrestrial broadcasting. However, since it can be used to provide a programme feed to cable networks and terrestrial (ground-based) transmitting stations, an explanation of the DBS technique is necessary. DBS poses technical problems very different from those encountered in conventional terrestrial television coverage because the received field strengths are much lower due to the low radiated power and the great path length. The radiated power is low because satellites rely on solar panels and rechargeable batteries for their power supply and the size of a transmitting antenna that can be launched into orbit is limited. The path length is long because it is necessary to use a geostationary orbit above the equator so that the satellite remains in a fixed position when viewed from earth.

The use of DBS to provide additional programmes, however, offers a number of advantages. It is possible to cover a much greater area than is practicable with terrestrial television, and 'shadowing' by natural and man-made obstacles is reduced so that there are fewer pockets of poor reception. Over the UK about 1000 stations of high, medium and low power are required to provide a reasonably complete UHF television coverage. On the other hand, a single satellite can provide equal or better coverage for a greater number of programmes.

10.2 Satellite transmission

10.2.1 The geostationary orbit

Because of the low received field strength, DBS reception requires a high-gain antenna. Clearly, it would be expensive and complicated to provide a receiving antenna that 'tracked' a satellite across the sky and produced no signal when the path to the satellite was shielded by the earth, as illustrated in Figure 10.1(a). Fortunately, if a satellite is placed at a height of 38 000 km above the equator, it will travel round the earth at the same speed as the earth's rotation and this is known as a 'geostationary' orbit. The geostationary orbit is illustrated in Figure 10.1(b) and it will be seen that the satellite, when viewed from the earth, remains in the same position in the sky so that a fixed receiving antenna can be used.

The fact that, to obtain a geostationary orbit, the satellite is very distant from the surface of the earth means, of course, that the signal is heavily attenuated even if the received field strength only diminishes as the inverse of the path distance. In fact, due to atmospheric conditions, the attenuation is greater than this. Another problem is that, although a satellite in geostationary orbit is always 'visible' from the earth, there are times when it is shielded from the sun, and the solar panels providing the transmitter power supply become inoperative. It is therefore necessary to provide rechargeable batteries fed by the solar panels to obtain a continuous supply of power for the transponders.

10.2.2 TVRO and CATV

DBS satellite transmissions cater for two broad programme distribution applications:

1. *TVRO (television receive only).* This is the reception of DBS transmissions by viewers using a domestic receiver/antenna installation.
2. *CATV (community antenna television).* This is the reception of DBS transmissions with a receiving installation that feeds a cable distribution network to supply programmes to the viewers. Also under this heading can be included the reception of DBS transmissions for re-transmission by terrestrial transmitters.

DBS services are carried by two types of satellite:

1. *Communication satellites.* These have a low transmitter power (a few watts) and spare channels are used for television transmissions. A large receiving antenna is required and the transmissions are most suited to CATV applications and the provision of programme feeds for terrestrial transmitters. Examples of these satellites are Intelsat and Eutelsat.

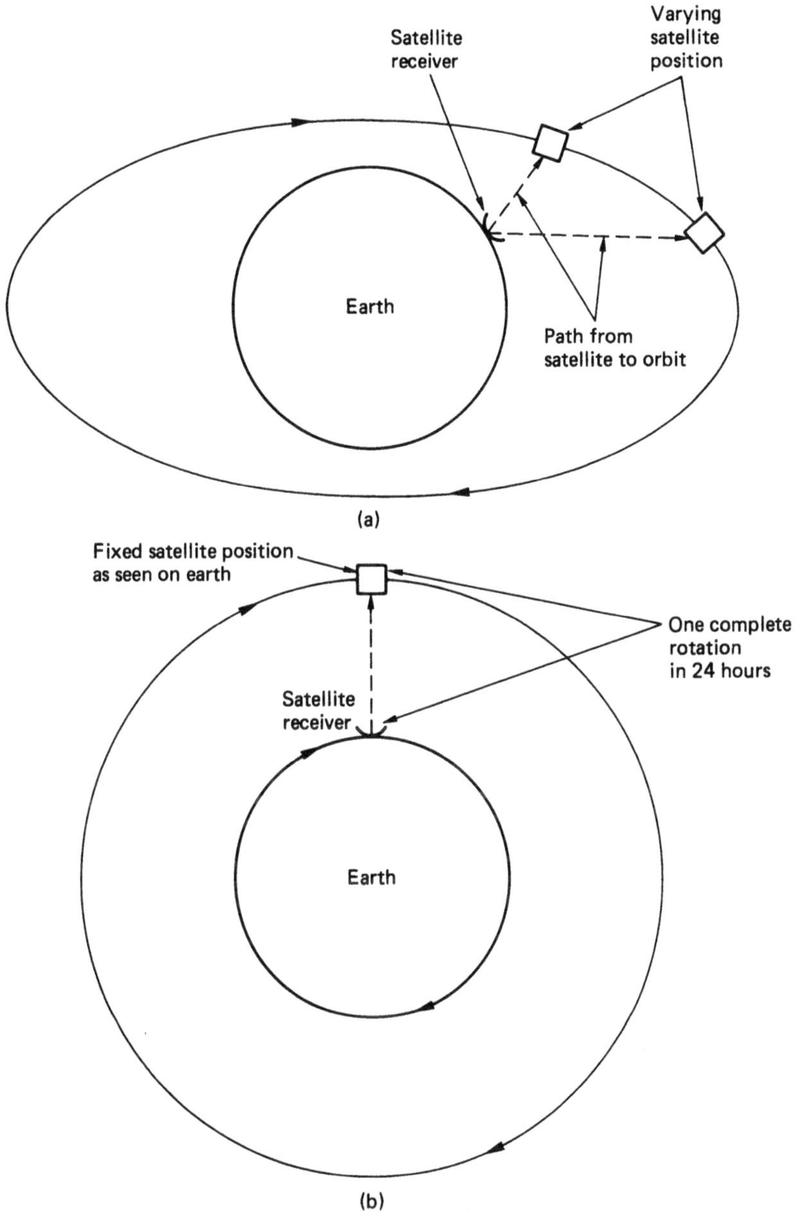

Figure 10.1 Satellite orbits. (a) Elliptical orbit. (b) Geostationary orbit

2. *TVRO satellites*. These have a medium–high-power output (from 50 to over 100 W) and are designed for DBS reception by a simple domestic installation with a small antenna. Examples are Astra (Sky TV and other services) and Marcopolo (BSB, British Sky Broadcasting) – now both BSkyB, after the amalgamation of Sky and BSB.

10.2.3 Satellite propagation

Transmission in the various frequency bands devoted to broadcasting relies on different propagation characteristics. In the LF and MF bands, the sky-wave is absorbed by the ionosphere during daylight hours and most transmissions therefore make use of the ground-wave. In the HF band (6–26 MHz) the ionosphere reflects transmissions by both day and night and it is therefore possible to communicate over distances of several thousand kilometres. In the VHF and UHF bands (50–1000 MHz) the reflecting effect of the ionosphere steadily diminishes so that reliable long-distance transmission cannot be achieved and services, once again, use the ground-wave. However, as the frequency is raised above 1000 MHz, the ground-wave suffers increasingly higher attenuation and, except for some short-range applications, terrestrial broadcasting becomes uneconomic.

At frequencies above 2 GHz (2000 MHz) the ionosphere becomes transparent and DBS therefore becomes possible. However, as the frequency is increased above 2 GHz the attenuation increases due to rain and other atmospheric moisture and places a limit of about 15 GHz on the maximum frequency that can be employed for DBS. Transmissions through space are attenuated much less than ground-waves and this, to some extent, offsets the very long path length inevitable with a satellite transmission.

10.2.4 Satellite frequency allocations

For acceptable television reception, the video signal-to-noise ratio (SNR) must be high and, because the attenuation of the signal is high, the received noise level must be as low as possible. Noise generated in receiving equipment can be minimized by careful design but radio noise is generated in space (this is known as 'sky noise') and can therefore be picked up by the receiving antenna of a DBS installation. Noise causes deterioration of the picture, and DBS therefore involves a struggle to reduce receiver noise and sky noise to an absolute minimum.

Although DBS transmissions are highly directional (the beamwidth of the transmitting antenna can be in the region of 1° or less), sidelobe radiation can cause interference over a wide area of the world's surface. For this reason, the World Administrative Radio Conference (WARC) has allocated shared frequency bands separated by 10 degrees of arc so that DBS satellite transmissions do not interfere with each other.

A satellite transmission, of course, requires two frequencies: an 'up' frequency at which the programme is sent up to the satellite and a 'down' frequency at which the signal is sent down for reception by DBS viewers and CATV ground stations. The down-link DBS frequency allocations are mostly in the 12-GHz region and the up-link bands are at somewhat higher frequencies.

10.2.5 Polarization

It has already been pointed out that for terrestrial television both VP (vertical polarization) and HP (horizontal polarization) are used. The use of different polarizations at neighbouring terrestrial transmitting stations reduces mutual interference and enables a greater number of stations to be sited in a given geographical area with a consequent improvement in the number of the population served. A similar situation applies in DBS, where the allocation of different polarizations enables the number of satellite programmes serving an area to be increased. It is, in fact, possible to transmit and receive adjacent channels from the same satellite if one is VP and one is HP.

During its passage through the earth's atmosphere the polarization of a DBS transmission is rotated. Thus, the receiving antenna needs to be rotated to avoid a loss of signal pick-up (this effect is known as 'depolarization loss'). It is possible, however, to use CP (circular polarization) for DBS transmission and in this case depolarization loss does not occur. For CP the satellite antenna transmits two equal-amplitude signals in phase quadrature, one of which is VP and the other HP, and the resulting signal has a polarization that rotates with time, as illustrated in Figure 10.2. The ground station receiving antenna is similar to the transmitting antenna in that it comprises two antennas disposed at right-angles and arranged to provide quadrature signals. The polarization of the transmitting antenna can rotate in either direction and this means that the receiving antenna must have the same sense of polarization rotation if it is to receive the transmitted signal. Thus with CP, as with VP and HP, there are two polarizations, referred to as RHC (right-handed circular polarization – clockwise rotation of the polarization when facing the antenna) and LHC (anticlockwise rotation). By using RHC and LHC it is possible to transmit

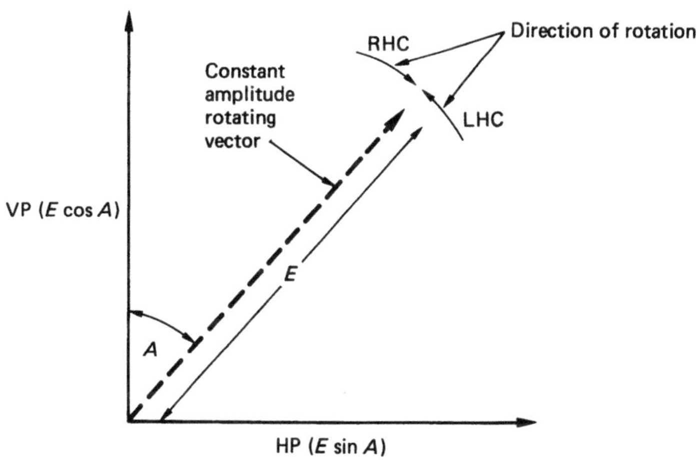

Figure 10.2 Circular polarization

adjacent channels from the same satellite just as with VP and HP. Further, it is obvious that, with CP, depolarization no longer causes a reduction. Most earlier satellites used VP and HP but many dedicated DBS satellites now use CP.

10.2.6 The satellite transponder

An important part of the satellite is the transponder that receives the signal from the up link, changes its frequency to that of the down link and amplifies its level as illustrated in Figure 10.3. The 'up' and 'down' frequencies, of course, need to be different to avoid instability due to pick-up of the transmitted signal by the satellite receiving antenna. It is quite difficult to amplify signals in the SHF range and most satellites make use of the TWT (travelling wave tube) illustrated in Figure 10.4. In the TWT, the signal passes along a helix in an evacuated glass tube about 60 cm long and a narrow stream of electrons is made to travel down the centre of the helix and interacts with the signal. Energy is drawn from the electron stream and transferred to the signal, thereby amplifying it. The TWT can provide output powers from a few watts up to more than 200 W with a gain in excess of 50 dB. This high gain is particularly important, as it simplifies the transponder which, as illustrated in Figure 10.3, may need to have only two amplifier stages.

The transmitted power of a satellite transmission is rated by the EIRP (effective isotropically radiated power) in dBW (dB relative to 1 W). This is the level of the radiated signal relative to that produced by a power of 1 W radiated by an isotropic antenna (an antenna that radiates equal power in all directions) and is given by:

$$\mathrm{EIRP} = 10 \log (W) + G$$

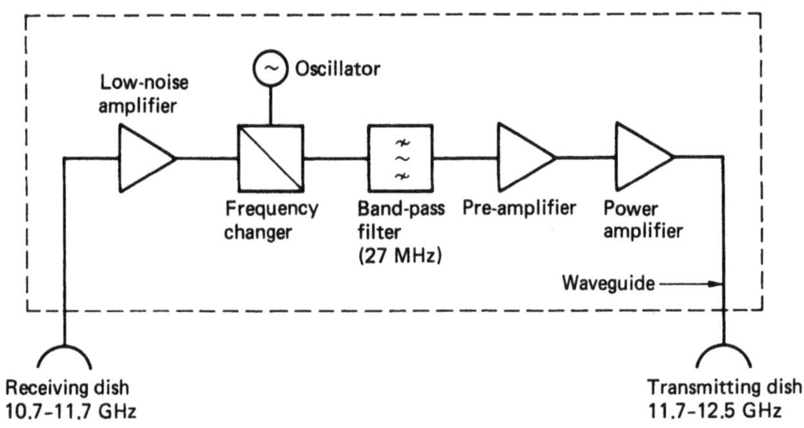

Figure 10.3 Satellite transponder

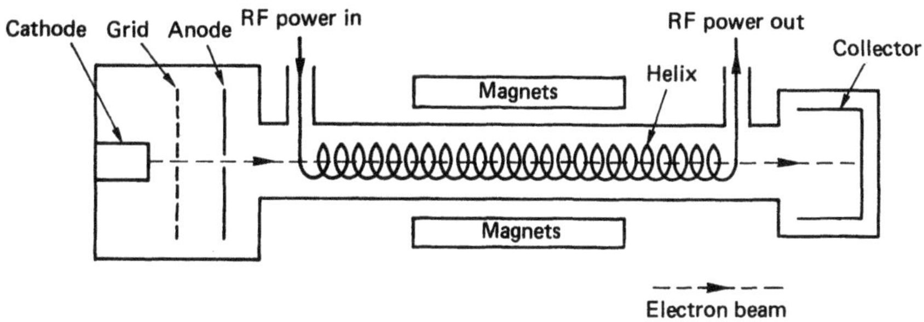

Figure 10.4 Travelling-wave tube

where *W* is the transponder power in watts, and *G* is the gain of the satellite antenna in dBi (see Appendix 2). For example, if the transponder has an output power of 50 W and the satellite antenna has a gain of 42 dBi the EIRP will be equal to 59 dBW (i.e. $10 \log 50 + 42$).

10.2.7 Bandwidth

The video bandwidth of the European PAL colour television transmissions on the 625-line, 50-fields system varies between 5 and 6 MHz and, taking account of the vestigial sideband and sound carrier, the complete signal, in each case, is contained in an 8-MHz RF channel width. For the American NTSC 525-line, 60-fields system the video bandwidth is 4.2 MHz and the channel width is 6 MHz. The video signal is amplitude modulated on the RF carrier and the channel width is therefore the minimum that will accommodate the complete vision plus vestige plus sound baseband signal.

The fact that DBS operates at the much higher frequencies in the SHF band means that wider RF bandwidths can be allocated without unduly limiting the number of channels available. The channels enable a signal bandwidth approaching 30 MHz to be used (26 or 27 MHz are commonly used in Europe and 24 MHz in the USA). Adjacent channel allocations overlap with a channel spacing in the region of 15 MHz (for example, the channel spacing is 14.58 MHz in the USA and 14.75 MHz for the 11.2–11.45-GHz band used for the Astra satellite in Europe). Adjacent channels are used either for the coverage of areas separated by about 10° (when viewed from the satellite) or for the same area using different polarizations (VP and HP or RHC and LHC).

10.3 Satellite reception

10.3.1 Modulation

As has been pointed out in Section 10.2.7, a bandwidth approaching 30 MHz is provided for each satellite transmission and in practice the transmitted bandwidth is 26–27 MHz. This relatively wide bandwidth enables the video and sound transmissions to be radiated on a carrier in the 12-GHz band by frequency modulation (FM), which results in an improvement in SNR after demodulation. When FM transmission is used, a (theoretically infinite) series of sidebands is generated and, to avoid distortion, the RF bandwidth needs to be equal to twice the maximum frequency deviation plus twice the maximum modulating frequency. For example, for European PAL transmissions the maximum video modulating frequency is 6 MHz and, allowing for the sound transmission and vestige, the total modulation bandwidth is about 7 MHz. Therefore, a maximum frequency deviation of 6.5 MHz will result in an RF bandwidth of $2(6.5+7)=27$ MHz. In North America the maximum video modulating frequency is 4.2 MHz and the total modulation bandwidth is therefore less than in Europe so that a 24-MHz channel width can be used.

The advent of DBS with its wider bandwidth and need for new receiving equipment has made it possible to introduce a new modulation system. The main defect in the PAL, SECAM and NTSC systems is the occurrence of 'cross-colour' and 'cross-luminance' which, respectively, show themselves as spurious colour patterns and luminance errors. These defects are due to the fact that, as explained in Chapter 7, the luminance (black and white) and chrominance (colour) signals share the same video bandwidth. Cross-colour occurs when luminance information in the chrominance band is mistaken by the receiver for chrominance, and cross-luminance when the chrominance band is mistaken by the receiver for luminance. The D-MAC system proposed by National Transcommunications Limited (previously the IBA – the UK Independent Broadcasting Authority) eradicates these errors and also caters for digital sound transmission and the eventual introduction of HDTV (high-definition television). In the MAC (multiplexed analogue components) system, cross-colour and cross-luminance are eliminated by transmitting the luminance and chrominance as separate signals, as illustrated in Figure 10.5. Figure 10.5(a) shows the transmitted signal on one line of the standard PAL system where the luminance and chrominance signals appear as a 'composite waveform'. Figure 10.5(b) shows the MAC line waveform with the chrominance and luminance signals appearing as separate 'packets'. With the MAC system it is, of course, necessary to 'hold' the chrominance information until the luminance information arrives but, with modern solid-state circuitry, this is quite easily achieved. However, the decision in 1990 that Sky TV and BSB should amalgamate and use only the Astra satellite (which radiates standard PAL) appears to put back the prospect of improved PAL and HDTV, though it has now been agreed that all new receivers will incorporate circuitry to receive D-MAC transmissions.

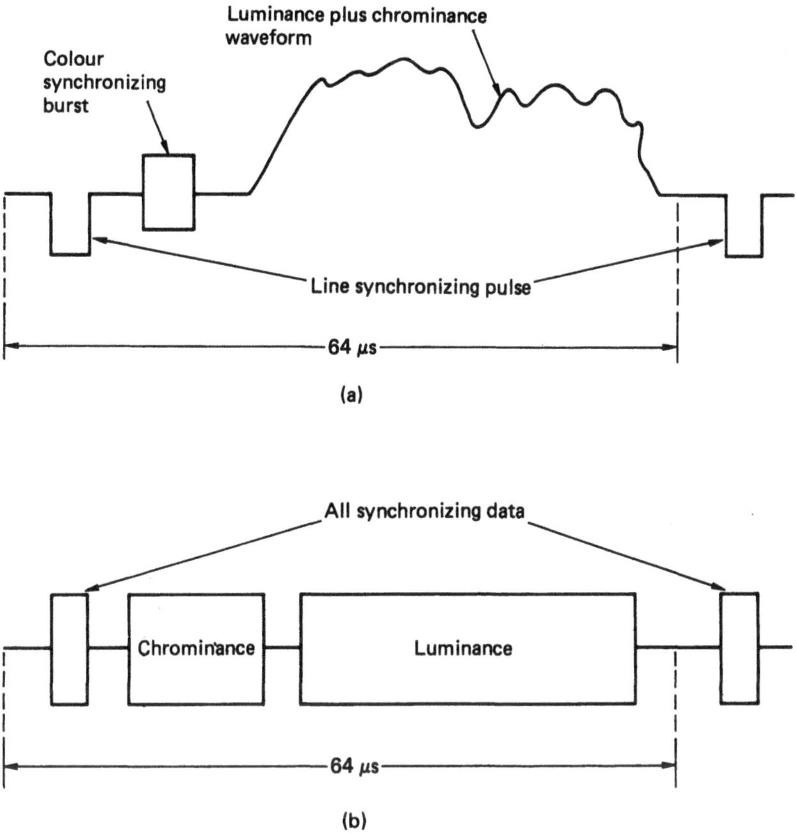

Figure 10.5 PAL and MAC line waveforms. (a) Standard PAL system. (b) MAC system

10.3.2 Signal-to-noise ratio

The long path length and the consequently low received field strength means that SNR is a parameter of prime importance in DBS. Obviously, every effort is made to radiate a signal with very low noise from the satellite and the received SNR is determined by noise from two sources:

1. Noise generated in the receiving equipment and, in particular, in the early stages.
2. Atmospheric and galactic noise arriving at the antenna. Atmospheric noise is due to natural electrical activity in the atmosphere and galactic noise is generated in space.

Signal power is easily calculated and, to determine the received SNR, some way of calculating noise power is required. There is more than one way of doing this but the method based on 'noise temperature' is probably the most straightforward. Any physical resistance generates noise and the 'available' noise power that can be delivered into a load is given by:

$$P = kTB \text{ watts}$$

where k is Boltzmann's constant (1.38×10^{-23} W Hz^{-1} K^{-1}), T is the temperature in °K, and B is the bandwidth in Hz. This means that a resistance of R at a temperature of T will deliver P watts of noise into a matched noiseless terminating resistance of R via a filter of bandwidth B.

Noise power thus depends on temperature and, for a given bandwidth, the amount of noise can be immediately determined if the 'noise temperature' is known. This is the temperature in kelvins to which a resistance must be raised to provide the amount of noise that actually occurs. This is a broad definition that can be used to assess the effect of all the noise sources in a communications system:

1. For receivers the noise temperature is that to which the input resistance must be raised to generate the actual output noise (this will include the noise generated in all the stages of the receiver).
2. For antennas the noise temperature is that to which a physical resistance equal to the antenna input impedance must be raised to generate the amount of noise actually picked up by the antenna.

The total noise temperature is given by the sum of the receiver and antenna noise temperatures and the total noise power is given by:

$$N_p = -168.6 + 10 \log (T_s b) \text{ dBW}$$

where T_s is the sum of the antenna and receiver noise temperatures, and b is the bandwidth in MHz. For example, if the antenna and receiver noise temperatures are 30 and 200 K respectively and the bandwidth is 27 MHz, the noise power is:

$$N_p = -168.6 + 10 \log [(30 + 200) \times 27] = -131 \text{ dBW}$$

Further, if the signal power is -120 dBW (a typical value), the SNR is:

$$\text{SNR} = -120 + 131 = 11 \text{ dB}$$

In DBS reception three SNRs are of interest:

1. The carrier-to-noise (C/N) ratio is the SNR referred to the receiver input. Since the RF signal is frequency modulated (i.e. it has a constant amplitude), C/N ratio is simply the RMS signal power divided by the noise power calculated as above.
2. The signal-to-noise (S/N) ratio is the SNR at the receiver video output. This is greater than the C/N ratio because there is an improvement in SNR resulting from the demodulation of the FM signal and because the signal power is defined as the peak-sync power and not the RMS power. The increase of S/N ratio over C/N ratio depends on the precise modulation parameters but is about 18 dB.

3. The effect of noise on a picture depends on its spectral distribution as well as on its power level. Thus, for a given value of S/N ratio the subjective picture impairment will vary with the spectral distribution. It is desirable, therefore, to have an objective measurement that will give the same value of SNR for different noise powers and spectral distributions having the same subjecting effect. This can be achieved by measuring the video noise via a low-pass network and this results in a further increase in SNR so that weighted S/N ratio is about 13 dB higher than unweighted S/N ratio.

In practice, a C/N ratio of 12–17 dB will give good to very good domestic reception and the corresponding unweighted and weighted S/N ratio values are 30–36 dB and 43–48 dB respectively. When SNR is discussed, it is obviously important to be sure whether C/N ratio or S/N ratio (unweighted) or S/N ratio (weighted) is referred to.

10.3.3 Satellite antennas

Antennas for DBS transmission and most antennas for DBS reception are of the parabolic dish type. The parabolic form is used as it has the property that, when it is used for reception, incoming rays are accurately directed to the focus as illustrated in Figure 10.6(a) (if a spherical dish is used this does not happen and there is a loss of gain as illustrated in Figure 10.6(b)). Conversely, when transmitting, the parabolic dish emits a parallel beam so that this shape has the optimum characteristics for both transmission and reception. Dish antennas have a gain that increases with the diameter but the size is limited in the case of both the transmission from the satellite and reception on the ground. In the case of transmission, the limitation is imposed by the fact that the dish has to be launched into orbit and it is difficult to do this with a diameter of more than 3–4 m. In the case of domestic reception the size of the dish is limited by cost and by the fact that it has to be suitable for mounting on the viewer's house. For these reasons, receiving dish diameters in the range of 30–90 cm are commonly used. The gain of a dish antenna operating in the 12-GHz band and with an efficiency of 60% is given, approximately, by:

$$G = 20 \log (D) + 40 \text{ dBi}$$

where D is the dish diameter in metres. For example, 60- and 90-cm dishes have gains of 36 and 39 dBi respectively.

Larger diameter antennas are used for transmission to satellites and for CATV systems where cost is a relatively less important consideration.

The efficiency of a dish receiving antenna is less than 100% because not all energy incident on it arrives at the focus (correspondingly, for a transmitting dish not all the energy supplied to the dish is focused into the transmitted beam). Loss of efficiency also occurs when the feed arrangement obstructs the path of the transmitted energy or, in the case of reception, shields the received energy

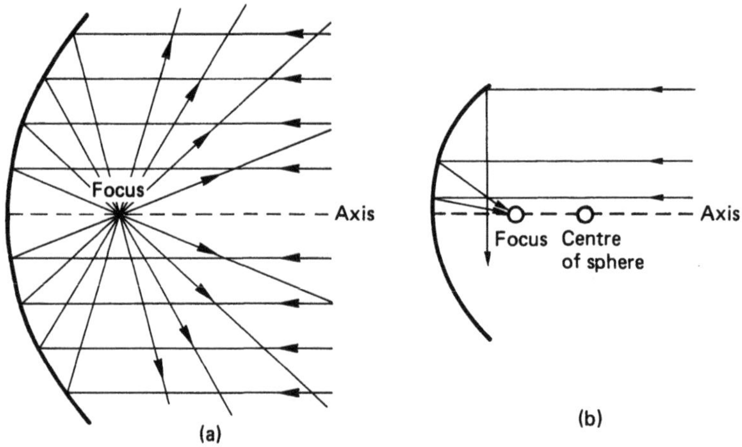

Figure 10.6 Dish antennas. (a) Parabolic dish. (b) Spherical dish

from the dish. A second important performance parameter of a receiving dish is the noise temperature and this is affected by both received sky noise and the energy picked up after reflection from the ground. There are four main types of dish antennas, illustrated in Figure 5.2, and their advantages and disadvantages can be summarized as follows.

(a) The front-fed (primary focus) type

This is the cheapest form of dish antenna. The disadvantages are that the feed arrangement partially shields the dish from the incoming signal with a resulting loss of gain, and sensitivity to ground-reflected signals leads to a relatively high noise temperature. The efficiency is in the region of 50%.

(b) The Cassegrain type

This is a more expensive type with a sub-reflector and the advantages are that the efficiency is higher and that the feed arrangement is relatively insensitive to ground-reflected signals, which results in a lower noise temperature. The sub-reflector shields the dish from the incoming signals but, with careful design, efficiencies up to 60% are possible. The fact that the feed arrangement is behind the dish makes the overhaul and maintenance easier in the case of large dishes.

(c) The offset type

This can be of the front-fed or Cassegrain type and the advantage is that shielding of the incoming signals by the feed arrangement is much reduced. The efficiency is therefore increased and can be as high as 70%. The Cassegrain offset dish is the more expensive and, as a result, many DBS receiving dishes are of the front-fed, offset type.

10.3.4 Amplifiers and converters

The 'front-end' amplifier of a DBS receiving system is often termed the 'low-noise block converter' (LNB) and to reduce losses and noise it is usually bolted directly onto the dish feed system. The LNB amplifies the weak incoming signals and then converts them to a lower frequency so that a cable, rather than an expensive waveguide, can be used to carry the signal to the satellite receiver. As shown in Figure 10.7, in the satellite receiver, the frequency-modulated radio frequency is demodulated to baseband and then remodulated at UHF to feed a standard PAL, NTSC or SECAM television receiver.

The LNB is used to amplify very low-level RF signals and must therefore have a low noise temperature (i.e. it must introduce as little noise as possible). The noise performance of an LNB is often referred to as the 'noise figure', which is related to the noise temperature by the relationship:

$$F = 10 \log [(T + 290)/290] \text{ dB}$$

where F is the noise figure, T is the noise temperature of the LNB in kelvins, and 290 is the ambient temperature in kelvins. For example, an LNB with a noise temperature of 150 K will have a noise figure of $F = 1.8$ dB.

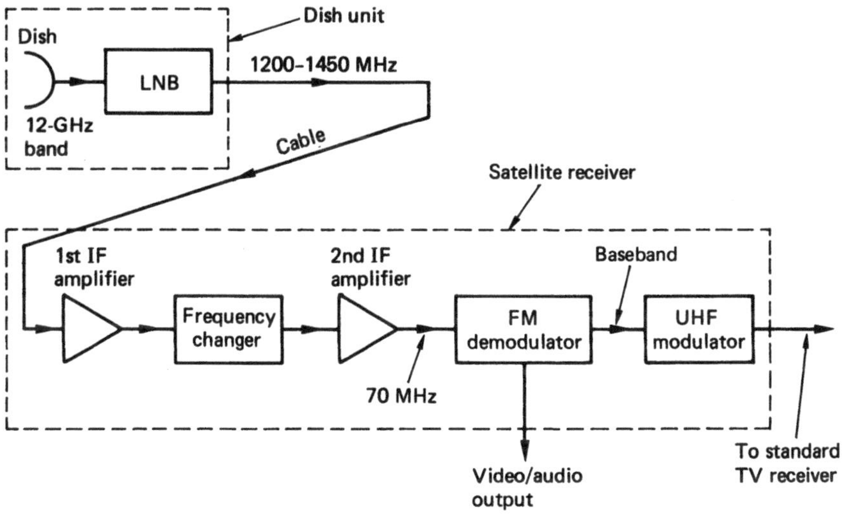

Figure 10.7 TVRO satellite receive arrangement

There are three basic types of low-noise amplifier for use in the SHF band:

1. *Parametric amplifiers* which operate by varying the capacitance of a varactor diode. For very low noise, the varactor needs to be cooled down to almost absolute zero and is therefore unsuitable for DBS applications.
2. *Tunnel diodes* which are fairly difficult to operate and have limited amplification. They are now seldom used for DBS applications.
3. *Gallium arsenide field-effect transistors (GaAsFETs)* which provide a high gain together with a low noise temperature (this can be as low, or lower, than 100 K). The GaAsFET has now superseded parametric and tunnel diodes for DBS applications.

Although the major part of the noise power at the output of an LNB is due to the first stage, the later stages also add a contribution. When the noise temperature (or noise figure) of an LNB is quoted it is referred to the input terminals but allows for all the noise. In other words the amplifier is assumed to be noiseless and to have a noise source at the input terminals such that the output noise power is that occurring in practice.

10.3.5 Noise budgets

The performance of a DBS receiving arrangement is measured by the SNR that is achieved and, as will be seen, this depends on a large number of factors. In calculating the SNR, it is therefore useful to draw up a 'noise budget', and an example is shown in Figure 10.8 for a typical case of DBS reception. Here the C/N ratio is such as to be well within the range for good reception. If the receiving installation were for providing a programme feed for a terrestrial transmitter or a cable network it would be necessary to legislate for a value of C/N in excess of 20 dB, which would require a receiving dish antenna with a diameter of 2 m or more.

10.3.6 The G/T parameter

In the particular case of DBS reception the path loss, frequency and EIRP are fixed and the parameters that can be varied by the system designer are the antenna gain and the noise performance of the receiving equipment. A convenient design parameter, which is expressed in dB, and is the antenna gain in dBi minus the total system noise temperature expressed in dB (i.e. $10 \log (T)$, where T is the sum of the receiver and antenna noise temperatures). It will be seen from Figure 10.8 that if G/T is held constant then the signal-to-noise-ratios C/N and S/N will remain constant. A value of G/T can therefore be chosen to provide the required SNR and the most cost-effective combination of antenna gain and equipment noise temperature established.

NOISE-POWER BUDGET

(a)	ERIP dBW	52
	Path loss dB	205
	Atmospheric loss dB	1
(b)	Total path loss dB	206
	Dish diameter cm (di)	90
	Dish efficiency % (ef)	60
	Frequency GHz (fr)	11.42
(c)	Dish gain dBi (Gain = 10 log (0.00011 × di² × fr² × ef))	38.4
(d)	O/P signal power dBW (a) − (b) + (c)	−115.6
(e)	Boltzmann's constant in dB	−228.6
	T_r (LNB noise temp.) K	200
	T_a (antenna noise temp.) K	50
	Total noise temp. $T_s = T_r + T_a$ K	250
(f)	Total noise in dB = 10 log (T_s)	24
	RF bandwidth MHz	26
(g)	Bandwidth in dB = 60 + 10 log (B)	74.2
(h)	Total noise power dBW (e) + (f) + (g)	−130.4
	C/N (carrier/noise ratio) dB (d) − (h)	14.8
(j)	Unweighted video S/N dB	32.8
(k)	Weighted video S/N dB	45.8

Figure 10.8 Typical noise budget

An example of the use of the G/T parameter can be based on the noise-power budget of Figure 10.8. Here the antenna gain 'G' is 38.4 dBi and the noise temperature 'T' (in dB) is 24 so that the G/T parameter is 14.4 dB. It is not possible to increase the antenna gain and reduce the noise temperature (or vice versa) so as to keep G/T constant. For example, it might be more cost-effective to use a 100-cm dish (giving an increase of gain of about 1 dB) and use an LNB with a noise temperature of 25 dB (an increase from 250 to 315 K).

10.4 Wide-screen high-definition television

The European 625/50 fields and North American 525/60 fields television standards have been in use for many years, during which time there have been many developments in the fields of signal origination, processing and transmission. There is general agreement amongst broadcasters that an improved picture standard should be introduced and this means that there should be an improvement in picture definition so that larger displays can be used to full advantage. In fact, the size of picture displays has increased steadily over the past few decades and further increases (including the application of solid-state techniques to 'flat' displays) will undoubtedly occur in the future.

The need for improved picture definition means that both the number of lines and the number of picture elements in a line (pixels) must be increased. It is also necessary to consider the aspect ratio of the transmitted picture (the ratio of picture width over picture height). From the commencement of television broadcasting an aspect ratio of 4:3 has been used, but there is now general agreement that this is not the optimum for home viewing. A wide-screen display (one with greater aspect ratio) will give an increased sense of realism for the viewer and for this reason many motion pictures are now made in this type of format. Summarizing, therefore, a new television standard needs greater resolution (i.e. a greater number of lines and a greater number of picture elements per line) and a larger aspect ratio. For use in Europe the proposed HDTV standard is:

Aspect ratio	16:9
Number of lines	1250 (interlaced)
Field frequency	50 Hz (picture frequency 25 Hz with two interlaced fields per picture)
Pixels per line	1920 (picture elements)

A new standard also needs to change the method of encoding the picture signals so that the cross-colour and cross-luminance distortions that occur with standard PAL and other transmission systems are eradicated. One way of doing this is to use the D-MAC system proposed by the UK Independent Broadcasting Authority (IBA) – now National Transcommunications Ltd – in which the chrominance and luminance signals are transmitted separated in time.

Without resorting to some form of bandwidth compression a television system on the HDTV standard would require a channel width in excess of 30 MHz as opposed to the 8 MHz in Europe and 6 MHz in North America used for terrestrial broadcasting. For DBS the channel width is 27 MHz but this is inadequate for HDTV. Thus, the introduction of television transmission on the HDTV standard involves two problems:

1. To compress the transmission into the channel widths currently available for DBS and ideally also for terrestrial broadcasting so that a high-resolution, 16:9 aspect ratio, picture can be received on a suitably designed receiver. This is the problem of HDTV.
2. To arrange that the transmissions are 'compatible' in the sense that they will give an acceptable picture on a standard European 625/50 or North American 525/60 receiver with a 4:3 aspect ratio. Clearly the picture in this case cannot be the equal of that received on an HDTV receiver and this is the problem of EDTV (enhanced-definition television).

10.4.1 HDTV (high-definition television)

It is easy with existing technology to generate a wide-screen, high-definition signal from a suitably equipped studio, VTR or film scanner. The problem is how to distribute this wideband signal to the viewer with minimal loss of quality through the restricted channel width of a DBS transmission. Bandwidth reduction systems were studied for many years without any practical outcome because the ability to store a complete field of a television picture was required and the equipment to do this was not available. Digital techniques have made large amounts of storage and comprehensive signal processing practicable and a number of methods of compressing an HDTV signal into a DBS bandwidth have been demonstrated.

One such system is based on the MAC system (separate transmission, displaced in time, of the luminance and chrominance signals) and is known as HDMAC (high-definition MAC). The bandwidth required for transmission of a picture increases as the amount of motion increases and this fact can be used as a basis for bandwidth compression. The system employs two separate coding paths, one of which has high resolution for relatively stationary areas (e.g. the background of a picture). The other path has reduced resolution for use in picture areas where there is motion. The decision as to which of the two paths is used for a particular area of the picture is taken in the encoder based on a comparison of the original wideband input signal and the reconstructed signals via each of the coding paths. By making the reconstruction circuits in the encoder identical to those used in the receiver it is possible to predict which of the two paths will give the better picture rendition. The coding decisions are signalled to the decoder in the receiver via a separate data channel using a technique known as DATV (digitally assisted television). The HDMAC-DATV system assumes the use of a separate RF transmission channel and does not cater for compatible reception on 4:3 aspect ratio PAL and NTSC terrestrial receivers. Such channels can be made available in the SHF bands used for DBS but not in the VHF and UHF bands used for terrestrial television, where there is no spare bandwidth available.

10.4.2 EDTV (enhanced-definition television)

It is obvious that it will be very desirable to improve the quality of the existing terrestrial transmissions in the face of competition from wide-screen HDTV DBS services. It is not possible to achieve the full advantages of HDTV within the restricted bandwidth of terrestrial transmissions but systems with a quality approaching that of HDTV are possible. Such systems are known as EDTV (enhanced-definition) systems and the objectives are:

1. Compatibility with standard receivers (i.e. an acceptable picture must be provided on a standard receiver with a 4:3 aspect display).

2. Reduction of PAL and NTSC defects.
3. Wide-screen display on an HDTV receiver.
4. Improved horizontal and vertical resolution (at least on HDTV receivers).

(a) Compatibility

Any modifications or additions to the transmitted signal must be such that they are 'transparent' (i.e. they are ignored by standard receivers). It would be quite impractical to expect millions of viewers to have their receivers modified or to adjust them when receiving EDTV transmissions.

(b) Reduction of PAL and NTSC defects

The well-known defects of cross-colour and cross-luminance arise because the luminance and chrominance components are interleaved to share the same spectrum. These defects limit the performance of present standard transmissions, particularly for large-screen displays. Some improvement can be achieved by using improved filtering when separating chrominance and luminance in receivers with large-screen displays. Also, some improvement can be obtained by changes in the encoding but, obviously, these must be transparent to standard receivers. A number of improved systems have been proposed, all of which exhibit improvements which outweigh any new form of distortion that may be introduced.

(c) Wide-screen display

An EDTV system must transmit a wide-screen picture for wide-screen, 16:9 aspect ratio receivers and there is obviously a difficulty here in displaying the picture on a standard receiver with a 4:3 aspect ratio. There are two methods of achieving this, which are known as 'letter box' and 'side panels'.

Letter box

Here the height of the transmitted picture is reduced by 25%, as illustrated in Figure 10.9(a), so that a 16:9 aspect ratio picture can be displayed on a standard receiver. This leaves a number of blank lines at the top and bottom of the picture that can be used for the transmission of additional data for the reconstruction of an EDTV picture on a wide-screen receiver. For the standard receiver, the technique has the advantage that it is the only way that the whole picture area of a 16:9 aspect ratio picture can be reproduced on a 4:3 aspect ratio display. The disadvantages are that the viewer, with a standard receiver, is

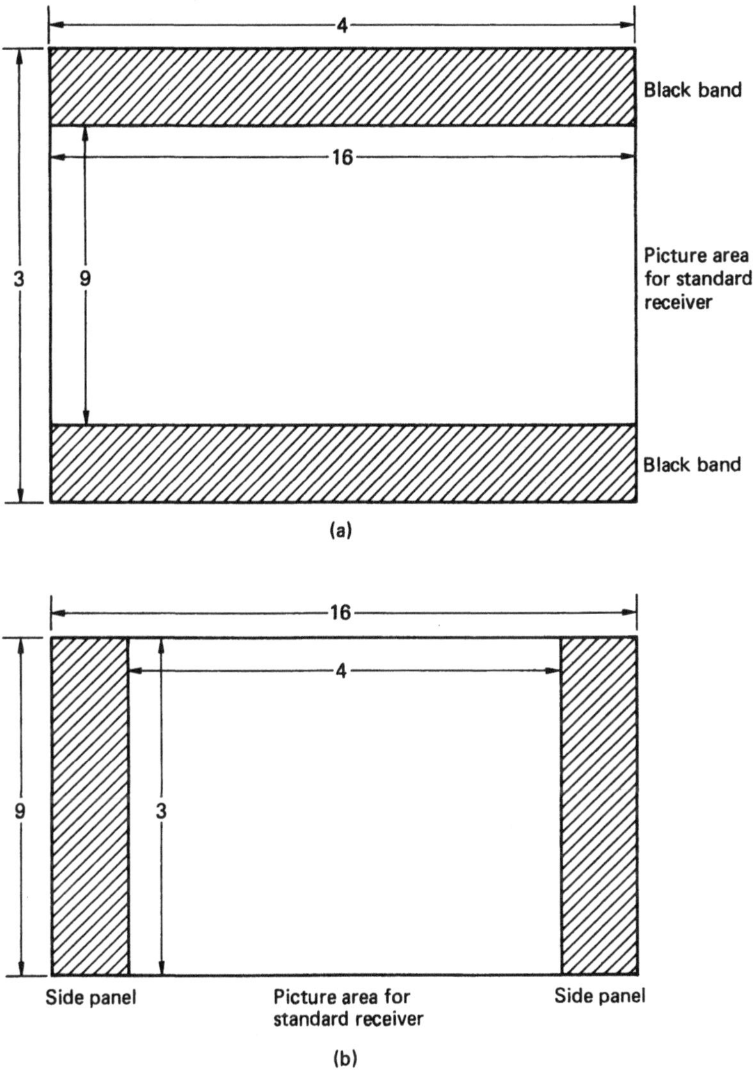

Figure 10.9 Compatible methods for standard PAL. (a) Letter box wide aspect ratio. (b) Side panels wide aspect ratio

unlikely to be willing to watch all programmes with black bands at the top and bottom of the screen. Further, the vertical resolution is reduced, making it more difficult to resolve fine detail (in titles, for example).

The additional data in the black bands used to construct a 16:9 aspect ratio in a wide-screen receiver must be transparent to a standard 4:3 aspect ratio receiver and this is not easy to arrange.

Side panels

An alternative to letter box is to transmit a standard 4:3 aspect ratio or a 4:3 portion of a 16:9 aspect ratio picture in the normal manner. In this way, standard receivers display normal (full-size) pictures with normal vertical and horizontal resolution. For 16:9 aspect ratio pictures, 'side panel' information is transmitted (transparent to 4:3 aspect ratio receivers) so that the full wide-screen picture can be reproduced on advanced receivers as illustrated in Figure 10.9(b). The information is recovered by the advanced receiver and the side panels are 'stitched' back on to the main signal prior to final display. The disadvantages are that the viewer with a standard receiver gets only part of the 16:9 aspect ratio picture and it is difficult to stitch back the side panels on to the 4:3 aspect ratio picture without the 'joins' becoming visible.

It is possible to design compromise techniques embodying both letter box and side panels. A letter box with a height reduction of about 12% would be more acceptable to the viewer but would require a similar percentage of additional information transmitted as side panels plus some extra line information to restore full vertical resolution.

Further improvements in both vertical and horizontal resolution are probably needed for large EDTV screen displays. Line doubling will reduce the visible line structure and may be sufficient for the vertical resolution.

10.4.3 Standards

There are a number of different proposals for the radiation of wide-screen pictures by DBS and by terrestrial transmission. It is possible, therefore, that transmissions to any given country may be made on a number of different standards. For example, one standard might be used for EDTV-compatible terrestrial transmissions and another for HDTV-non-compatible transmissions by DBS. Further, the standards adopted by different countries may differ (quite apart from the difference in the present PAL, SECAM and NTSC systems in Europe and North America). It seems, therefore, that in addition to a number of transmission standards in different parts of the world, there may be two or more transmission standards in any one country. A few years ago this would have been an impossible situation, but with recent advances in solid-state techniques it is possible to manufacture receivers with a number of decoders for the reception of different standards. Thus, in the home, there might be a wide-screen, multi-decoder receiver for HDTV and EDTV transmissions and one or more standard receivers for reception of the present 4:3 aspect ratio transmissions and compatible (4:3 aspect ratio) versions of wide-screen (16:9 aspect ratio) transmissions.

Chapter 11

Feeders, transmission lines and waveguides

11.1 Introduction

In all broadcast applications the locations of the transmitter and antenna are separated (usually by at least 30 m) and it is necessary to connect them together; it is convenient to describe the means of connection as a 'feeder'. It is therefore important for the broadcaster planning a new installation to have some knowledge of the advantages, disadvantages and characteristics of the various types of feeder that are used in the various broadcast bands. Feeders are classified as being screened or unscreened. The commonest types of the former comprise waveguides, coaxial transmission lines and screened balanced pair transmission lines, and the later open-wire balanced and cage-type unbalanced transmission lines. Open-wire lines are used at frequencies up to about 30 MHz (the upper end of the HF band), coaxial transmission lines at frequencies up to about 1 GHz (the upper end of the UHF band) and waveguides mainly at frequencies above 1 GHz (the SHF or microwave band).

11.2 Transmission line and waveguide theory

Transmission lines use the transverse electromagnetic (TEM) mode in which the directions of the electric and magnetic fields are at right-angles to each other and at right-angles to the direction of propagation, as illustrated in Figure 11.1(a). A TEM wave can only be sustained on a transmission line when there are two conductors and these may be in the balanced or coaxial configuration as shown in Figure 11.1(b). A transmission line will only operate satisfactorily and with low loss when the spacing between the conductors does not exceed a small part of a wavelength at the transmitted frequency. For example, for a coaxial transmission line, the distance between the inner and outer conductors must be small compared with a wavelength and this places an upper limit on the operating frequency which is given approximately by $F = 5320/D$, where D is the outer diameter in inches and F the frequency in megahertz. For example, a six-inch diameter transmission line will have a maximum operating frequency of about 900 MHz for straight runs, but only about 800 MHz around 90° bends. Thus, with increasing frequency, the maximum size of transmission line that can

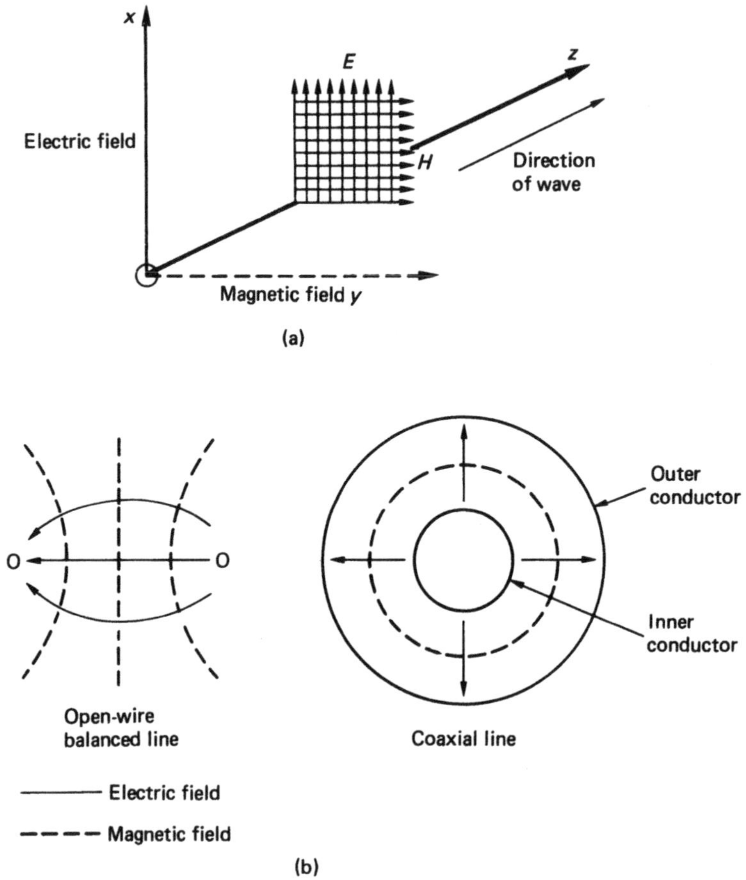

Figure 11.1 The TEM (transverse electromagnetic) mode. (a) TEM field directions. (b) Fields of transmission lines

be used decreases with a consequent decrease in power rating and increase in attentuation. Further, since for a given size the attenuation increases with frequency, transmission lines become impractical at frequencies above about 1 GHz (1000 MHz) except for short lengths and low powers. However, it is worth noting, and occasionally of practical importance, that for a given outer diameter the maximum operating frequency increases with increasing impedance.

In waveguides, unlike transmission lines, propagation can only take place above a certain frequency (known as the cut-off frequency). The cut-off frequency occurs when the cross-sectional dimensions become too small to sustain a mode of propagation in which the transmitted wave is reflected from one side of the guide to the other, as illustrated in Figure 11.2 for a waveguide with rectangular cross-section. Above the cut-off frequency the attenuation of a waveguide decreases until a frequency is reached at which unwanted modes,

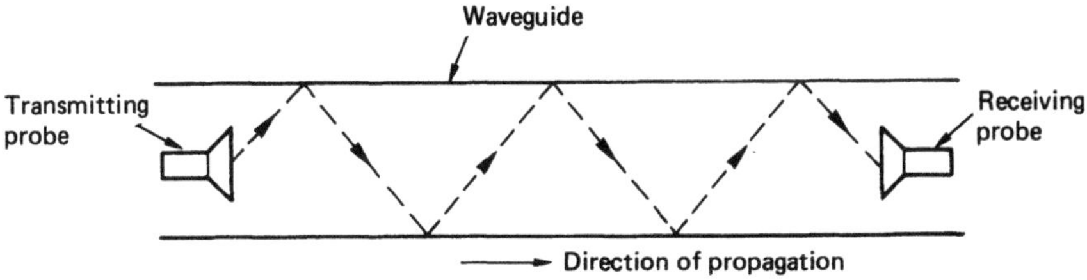

Figure 11.2 Propagation in a waveguide

which propagate at slower speeds, are generated and result in a sharp and prohibitive increase in attenuation. A waveguide of a given size therefore has a frequency pass-band rather like that of a band-pass filter. Because waveguide sizes decrease with increasing frequency they are well suited to use in the microwave band above 1 GHz but less attractive at lower frequencies on account of their large physical size. For example, a rectangular waveguide operating at a frequency of 200 MHz would have cross-sectional dimensions of approximately 0.9×0.45 m.

11.2.1 Characteristic impedance

Characteristic impedance is a parameter that is relevant to transmission lines but not (at least directly) to waveguides. The characteristic impedance of a transmission line is the impedance that would be seen at the input of a line of infinite length. Thus, if a finite length of line is terminated in its characteristic impedance, the input impedance will be equal to the characteristic impedance, as illustrated in Figure 11.3. This is important, as it means that if a line is so terminated there will be no reflection of energy back to the input and the loss in the line will be a minimum. For transmission lines with small losses (as is the case in broadcast transmission practice) the reactive term is very small and the characteristic impedance can be regarded as purely resistive. For a balanced, air-spaced, two-wire, open-wire line the Z_0 (characteristic impedance) is given by:

$$Z_0 = 276 \log (S/R)\ \Omega$$

where S is the spacing between the conductors, and R is the radius of the conductors. The value of Z_0 changes slowly with the ratio S/R and in practice is around 600 Ω. For example, a conductor radius of 1 cm with a spacing of 150 cm gives $Z_0 = 601\ \Omega$. To increase power-carrying capacity, four-wire lines are often used and the characteristic impedance is then approximately half that of a two-wire line (i.e. about 300 Ω).

The characteristic impedance of a coaxial line is given by:

$$Z_0 = 138E^{-1/2} \log (R_0/R_i)\ \Omega$$

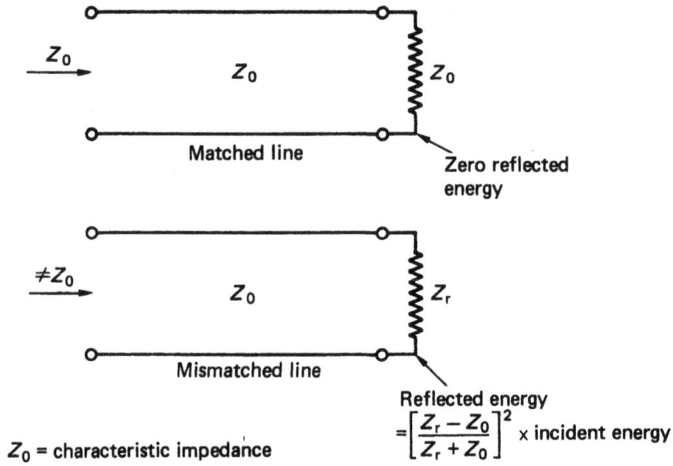

Figure 11.3 Characteristic impedance

where R_o is the radius of the outer conductor, R_i is the radius of the inner conductor, and E is the dielectric constant of the insulation (unity for air). As in the case of balanced lines, Z_o changes slowly with the ratio R_o/R_i and in practice is in the range 30–100 Ω. For lower values there is very little space between the inner and outer conductors and it is difficult to achieve uniformity. For higher values the radius of the inner conductor becomes too small to be practicable.

11.2.2 Attenuation

(a) Transmission lines

At DC (direct current) the current in a conductor flows uniformly over the cross-section. However, at high frequencies the current is forced to the outer surface and concentrated in a thin skin (this is known as the 'skin effect'). As a result of the reduction in the cross-sectional area carrying the current, the effective resistance is increased as compared with that at DC. The depth of the skin decreases with increasing frequency and the attenuation (expressed in dB) increases approximately as the square root of the frequency. For example, if the attenuation of 100 m of transmission line is 3 dB at 50 MHz, then the attenuation at 100 MHz will be about 4.2 dB.

(b) Waveguides

The simplest type of waveguide, from the theoretical point of view, is the rectangular form shown in Figure 11.4. A waveguide will transmit energy when distribution of the electric and magnetic fields is in a configuration known as a 'mode'. Modes are denoted by TE and TM, dependent on whether the direction of the electric or magnetic field is at right-angles to the direction of propagation. A rectangular waveguide is usually operated in the TE_{10} mode, which occurs when the greater of its lateral dimensions is greater than a half-wavelength and the smaller less than a half-wavelength. The transmitted energy will then 'bounce' from side to side as it travels from one end to the other. When the largest lateral dimension is equal to, or less than, a half-wavelength it is impossible to transmit energy and the corresponding frequency is therefore the 'cut-off frequency'. Above the cut-off frequency, the attenuation decreases because the number of reflections from one side of the waveguide to the other decreases, with a corresponding decrease in resistive loss. The attenuation decreases until the frequency is about twice the cut-off frequency, when the attenuation increases sharply as other, more complex transmission modes become possible.

The rectangular type of waveguide is made in relatively short, rigid lengths and has the advantage that quite sharp right-angle bends can be provided. However, it is not well suited to long runs as it is expensive and the joints between sections lead to impedance irregularities. Elliptical waveguides are in common use and have the advantage that they can be manufactured in long, continuous lengths which can be bent (a bending radius of 1 m or less is possible) so that a continuous run from the transmitter to the antenna can be accommodated with minimal need for joints and special angle sections. In the case of rectangular waveguides there is a theoretical operating bandwidth of about 2:1 but, in practice, the frequencies close to the low-frequency cut-off and the high-frequency mode change cannot be used so that the useful bandwidth is about 1.5:1. For elliptical waveguides, the operating frequency range is restricted to about 1.22:1 in order to prevent the generation of unwanted modes. It will be realized that the size decreases as the frequency increases and, for example, the following are typical cross-sectional sizes for elliptical waveguides (the two dimensions are the major and minor axes of the elliptical cross-section):

145 mm × 83 mm for 1.7–2.3-GHz operating band
29 mm × 19 mm for 12.2–13.25-GHz operating band

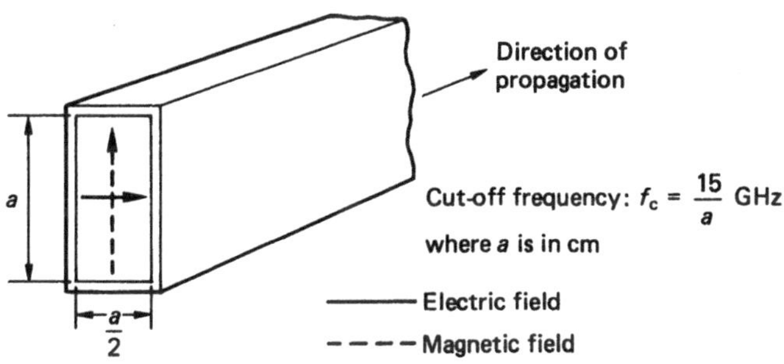

Figure 11.4 Rectangular waveguide (TE_{10} mode)

The attenuations of these waveguides decrease with frequency and for a 100-m length are:

145 mm × 83 mm: 1.5 dB falling to 1 dB
29 mm × 19 mm: 11.4 dB falling to 11.2 dB

The decrease in waveguide size with increasing frequency means that the power-carrying capacity decreases with increasing frequency. For example, the larger waveguide above has a power rating of 50 kW and the smaller 1 kW. This decrease of power rating does not result in any serious limitation in practice, as operating powers tend to be lower in the higher frequency bands.

11.2.3 Reflection coefficient and VSWR

It has already been pointed out that, if a transmission line is terminated in its Z_0 (characteristic impedance), the input impedance will be equal to the Z_0 and there will be no reflection of energy from the termination. In practice, however, although feeders are usually operated as closely as possible under this matched condition, irregularities in the transmission line or waveguide cause reflections and result in a deviation of input impedance from a value equal to the Z_0. The 'reflection coefficient' sums up all the impedance variations within a feeder and is the ratio between the vector sum of all the reflections and the level of the incident signal (measured at the sending end of the feeder). When assessing transmission line performance, it is common to use the term VSWR (voltage-standing-wave ratio), which is given by:

$$\text{VSWR} = (1 + r/100)/(1 - r/100)$$

where r is the reflection coefficient expressed as a percentage. For a length of line, terminated in its nominal Z_0, the VSWR measured at its input is known as the 'insertion VSWR' and is an important performance parameter.

VSWR is, in effect, the ratio of maximum to minimum voltage along a transmission line and, particularly when it is operated with a mismatched termination, has a significant effect on the power-handling capacity.

Instead of the reflection coefficient, the term 'return loss' is sometimes used (particularly in connection with waveguides) and is the ratio of the reflected signal (due to discontinuities and any terminating mismatch) and the input signal expressed in dB:

$$A = 20 \log (100/r) \text{ dB}$$

where A is the return loss in dB, and r is the reflection coefficient expressed as a percentage.

11.2.4 Power rating

As has been pointed out, limitation of power rating is seldom a problem encountered when operating waveguides. The power ratings of waveguides are relatively high and the transmitted powers are usually low. On the other hand, power rating is a very important factor in the case of transmission lines, particularly at frequencies in the VHF and UHF bands and above. In the case of coaxial lines, most manufacturers give two 'standard' power ratings defining the maximum peak and average powers that can be transmitted. The 'standard' mean power rating is the transmitted power at which (under standard operating conditions) the inner conductor reaches the safe working temperature for the dielectric, and the peak power rating depends on the maximum voltage that the transmission line can safely withstand. It should be noted that the mean power rating is dependent on, and the standard peak power rating is independent of, frequency. Most transmission line manufacturers give the peak and mean power ratings assuming an ambient temperature of 40°C, a matched load and dry air at normal atmospheric pressure between the conductors. For a given application it is necessary, therefore, to determine the peak and mean standard power ratings, which will then enable the type of line required to carry the power under the actual operating conditions to be selected. These ratings are obtained by multiplying the transmitter power by a number of 'operating factors' to obtain the equivalent standard powers that the line must be able to carry.

(a) Transmitter power

The peak and mean powers of a transmitter depend on the type of modulation. The operating factors w_p and w_m are given in Table 11.1.

Table 11.1

Modulation	*Reference power*	*Peak power factor* (w_p)	*Mean power factor* (w_m)
AM	Carrier	4.0	1.5
FM	Carrier	1.0	1.0
TV	Peak vision sync (and 1/10 power FM sound carrier)	1.73	0.73

For example, an AM transmitter with 50-kW carrier has a peak power of 200 kW and a mean power of 75 kW under maximum modulation conditions. A TV transmitter with a vision power of 50 kW will have a peak power of 85 kW and a maximum mean power of 36.5 kW (these figures include the contribution of the sound transmitter, assumed to be frequency modulated and of 1/10 the peak-sync power).

(b) VSWR

An increase in VSWR will decrease the peak and mean power-carrying capacities of a transmission line so that a higher standard power rating is required. The operating factors are:

$$v_p \text{ (peak power)} = V$$
$$v_m \text{ (mean power)} = V$$

where V is the VSWR at the transmission line input.

(c) Ambient temperature

Increased ambient temperature reduces the mean power-carrying capacity of a coaxial transmission line so that a higher standard rating is required. The operating factors are:

$$t_p \text{ (peak power)} = 1 \text{ (at all temperatures)}$$
$$t_m \text{ (mean power): as in Table 11.2}$$

Table 11.2

Transmission line type	*Ambient temperature (°C)*				
	20	30	40	50	60
BAS	1.0	1.0	1.0	1.0	1.0
SAS	0.8	0.9	1.0	1.1	1.18

BAS, balanced air-spaced;
SAS, coaxial semi-air-spaced with dielectric spacers or helix.

To simplify calculations, ambient temperatures of 20 and 40°C are sometimes taken to correspond, respectively, to temperate and tropical climates.

(d) Direct solar radiation

Heating due to direct solar radiation reduces the mean power-carrying capacity of a coaxial line so that a higher standard rating is required. The operating factors are:

$$r_p \text{ (peak power)} = 1 \text{ (under all conditions)}$$
$$r_m \text{ (mean power): as in Table 11.3}$$

Table 11.3

Line type	*Temperate zone*	*Tropical zone*
BAS	1.0	1.0
SAS	1.03	1.06

(e) Altitude

The peak power-carrying capacity of a transmission line at atmospheric pressure decreases with increasing altitude above mean sea level so that a higher standard rating is required. The operating factors are:

a_p (peak power): as in Table 11.4
a_m (mean power) = 1 (independent of altitude)

Table 11.4

Altitude above mean sea level (m)	0	250	500	750	1000	1250
Operating factor	1.0	1.02	1.06	1.1	1.14	1.18

(f) Pressurization

If a transmission line is pressurized (this, of course, applies only to coaxial cables using dielectric spacers or helix) the peak and mean power-carrying capacities are increased and a lower standard rating can be used. The operating factors are given in Table 11.5.

Table 11.5

	*Pressure**				
Factor	1	1.5	2	3	4
p_p (peak power)	1.0	0.6	0.4	0.25	0.17
p_m (mean power)	1.0	0.9	0.81	0.73	0.67

*The pressure is in bars (1 bar corresponds to atmospheric pressure at mean sea level). Operating pressures are often given in terms of excess pressure (i.e. the excess of pressure over atmospheric pressure). For example, 1.5 bars in the table corresponds to an excess pressure of 0.5 bars.

Note that if a cable is pressurized the altitude operating factors (a) must be taken as equal to unity.

(g) Buried coaxial lines

If part of a coaxial line is buried, or run in a duct, the operating temperature will increase and the mean power-carrying capacity will be reduced so that a higher standard rating will be required. The operating factors are:

$$b_p \text{ (peak power)} = 1$$
$$b_m \text{ (mean power)} = 1.05$$

The required standard ratings for the transmission line to be used are obtained by multiplying the reference transmitter power (see Table 11.1) by the product of w_p, v_p, t_p, r_p, a_p, p_p and b_p to obtain the peak power rating and by w_m, v_m, t_m, r_m, a_m, p_m and b_m for the mean power rating.

11.3 Transmission line and waveguide types

11.3.1 Coaxial transmission lines

(a) SAS (semi-air-spaced) transmission lines

SAS transmission lines are constructed with thin-walled tubular, inner and outer conductors which, in some designs, are corrugated to facilitate bending and reduce the minimum bending radius. The inner conductor is invariably copper and the outer conductor may be copper or aluminium. Typical applications are as medium–high-power transmission lines for MF, HF, VHFand UHF transmitters where these transmission lines have a number of advantages in that they:

1. Are 'semi-flexible' (the minimum bending radius depends on the diameter and ranges from a few centimetres to a few metres)
2. Have low attenuation
3. Are reliable
4. Are electrically uniform (the insertion VSWR is low)
5. Have a high power rating
6. Are totally screened, thus virtually eliminating coupling with adjacent lines and radiation hazards
7. Can be pressurized to increase the power rating.

SAS transmission lines have a mainly air dielectric with only sufficient dielectric material to ensure that the inner conductor is centred reliably and accurately. As a result, for a given size of outer conductor, the attenuation is low and the power rating is high. The characteristic impedance is normally 50 Ω but, in a few cases, 75 Ω (where the attenuation for a given outer diameter is slightly lower) is used. At one time, a 60-Ω characteristic impedance was introduced as a compromise between 50 Ω and 75 Ω but did not prove popular and is not now, normally, manufactured.

Typical standard power ratings and attenuations of 50-Ω SAS cables at 40°C and 100 MHz are as given in Table 11.6.

Table 11.6

Cable size (in)	*Mean power rating (kW)*	*Peak power rating (kW)*	*Attenuation (dB/100 m)*
$\frac{5}{8}$	4.0	32	2.0
$1\frac{1}{8}$	10.0	140	0.9
$1\frac{5}{8}$	15.0	230	0.6
$3\frac{1}{8}$	45.0	940	0.4
$6\frac{1}{8}$	140.0	2900	0.2
9	620.0	5900	0.15
12	830.0	9000	0.1

The performance at any other frequency can be estimated by multiplying the power and dividing the attenuation by a factor K, where K is equal to the square root of $100/f$ and f is the frequency in MHz. For example, the mean power rating and attenuation of $3\frac{1}{8}$-inch cable at 20 MHz will be approximately 100 kW and 0.16 dB/100 m.

The figures in Table 11.6 are approximate and intended for rough estimates of cable size and cost when a station is in the early planning stages. When transmission lines are ordered, therefore, the manufacturers should be consulted as regards up-to-date performance figures. It should also be remembered that, in practice, the factors described in Section 11.2.4 regarding environmental conditions must be taken into account.

SAS transmission lines are easy to install, require no routine maintenance and are highly reliable. However, if they do sustain accidental or deliberate damage, repairs require specialist staff and specialized equipment.

(b) Rigid coaxial transmission lines

Rigid coaxial transmission lines are assembled in short sections with anchor bullet assemblies for the inner conductor connection and flanges for the outer connections, with sealing rings for use out-of-doors where pressurization is necessary to prevent ingress of moisture. For long runs and high powers, expansion joints are essential. When compared with SAS cables, rigid lines may have the advantages of lower cost and that they can be repaired by non-specialist staff using simple equipment, but in practice installation errors occur and expansion joints often give trouble unless carefully maintained. Rigid lines tend to introduce a higher insertion VSWR than SAS lines due to discontinuities introduced by connections between the sections, uniformly spaced insulators, and the presence of expansion joints. For some applications, particularly in the higher frequency bands, to avoid resonance effects, it is necessary to specify the operating frequency(s) and so ensure that it (they) are not simply related to the regular spacing of any discrete insulators. As a result, they are most suited to high-power operation in the MF and HF bands, though some rigid transmission line components are inevitably needed for VHF and UHF installations. For these applications, and particularly at UHF, when transmitters are operated in parallel the rigid transmission lines carrying the power to the paralleling point must be generally similar, i.e. though they may follow different routes, they must include the same length of rigid line and number of bends to ensure similar electrical lengths over the frequency band of interest. Like SAS lines they can be buried in ducts and are thus less prone to accidental or deliberate damage. The peak and mean power ratings and attenuation for a given outer diameter are similar to those given in Table 11.6.

11.3.2 Screened balanced pair transmission lines

These types of lines are essentially two-wire lines inside rectangular screening cases and were indeed developed from two-wire open lines, to eliminate radiation hazards resulting from the use of such lines in transmitter halls and matrix switching complexes when high powers were involved. In addition to eliminating radiation hazards, unwanted coupling to other lines, particularly those in close proximity in switching matrices, is very much reduced, though still higher than between coaxial lines.

For these reasons this type of line is often used at high-power HF stations as an alternative to more costly but very high reliability coaxial lines.

11.3.3 Foam dielectric transmission lines

In 'foam dielectric' lines the space between the inner and outer conductors is completely filled with foam polyethylene, which prevents the ingress of moisture. These cables are much more flexible than the SAS type and are therefore suitable for installation where space is restricted and relatively sharp bends are necessary. The need to provide a really flexible cable restricts the maximum cable diameter to about $1\frac{5}{8}$ inches and the use of a foam dielectric restricts the power rating to less than that of a comparable SAS cable, though pressurization is not required. Foam dielectric cables are therefore often used for low-power unstaffed transmitter installations and Table 11.7 shows the power rating and attenuation at 100 MHz for a typical range of outer diameters. The characteristic impedance is normally 50 Ω, as in the case of SAS cables.

As in the case of SAS cables, the performance at any other frequency can be estimated by multiplying the power and dividing the attenuation by the factor *K*. For example, the power rating and attenuation of $1\frac{5}{8}$-inch cable at 20 MHz will be 29 kW and 0.3 dB/100 m.

Table 11.7

Cable size (in)	*Mean power rating (kW)*	*Attenuation (dB/100 m)*
$\frac{1}{4}$	1.4	4.3
$\frac{3}{8}$	1.8	3.0
$\frac{1}{2}$	2.9	2.1
$\frac{7}{8}$	6.6	1.2
$1\frac{5}{8}$	13.0	0.7

As in the case of Table 11.6, figures in Table 11.7 are approximate and intended for preliminary planning purposes. It should also be remembered that, in practice, the factors described in Section 11.2.4 regarding environmental conditions must be taken into account.

11.3.4 Open-wire transmission lines

Open-wire transmission lines are suitable for LF, MF and many HF high-power applications (above about 100 kW) and are less expensive than the coaxial or screened balanced pair types. However, their lack of screening, particularly for balanced lines, can result in significant coupling to adjacent lines (frequently giving rise to intermodulation products) and radiation hazards, so that at the highest power levels (over 300–500 kW) coaxial or screened balanced pair transmission lines are to be preferred and may be essential to avoid radiation

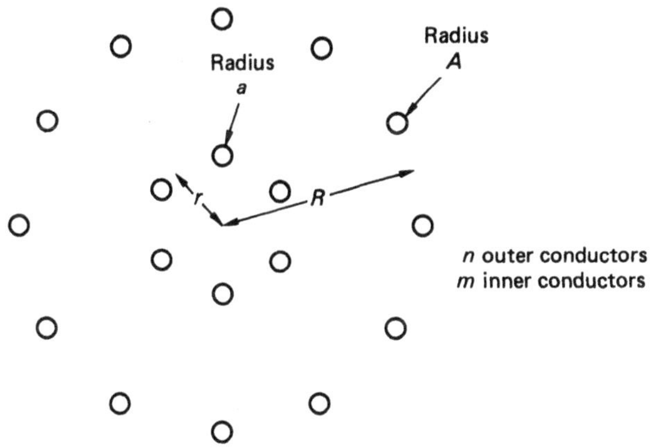

$$Z_0 = 138\left[\log\left(\frac{R}{r}\right) + \frac{1}{n}\log\left(\frac{R}{nA}\right) + \frac{1}{m}\log\left(\frac{r}{ma}\right)\right]\Omega$$

Figure 11.5 Cage-type, open-wire, unbalanced transmission line

hazards. The variation of characteristic impedance, inevitable with open-wire lines, leads to unacceptably high insertion VSWRs at frequencies above about 30 MHz.

Open-wire lines are of two main types:

1. The cage-type, open-wire coaxial line which is illustrated in Figure 11.5 is suitable for driving LF/MF antennas which, invariably, have an unbalanced input impedance. This type of line, which is effectively a skeletonized coaxial transmission line and hence partially screened, is simple to build and maintain and is less expensive than an SAS coaxial line for high transmitter powers (in the range 100 kW to 1 MW and above). The characteristic impedance of cage-type coaxial lines is higher than that of SAS lines and is in the range 120–250 Ω. Since they cannot be buried, cage-type lines are susceptible to accidental or deliberate damage but this disadvantage is offset by the fact that repairs can be made quite quickly by non-specialist staff using simple equipment.
2. The four-wire, open-wire balanced line which is illustrated in Figure 11.6 has a characteristic impedance of 300 Ω and is suitable, with the limitations previously mentioned, for driving HF antennas which, almost invariably, have a balanced input impedance. Like the cage type, the balanced line is simple to build and maintain and is less expensive than an SAS coaxial line for high HF transmitter powers (in the range 100–500 kW and above). Also, like cage-type lines, balanced, open-wire lines are susceptible to accidental or deliberate damage but can be repaired by non-specialist staff using simple equipment. A 300-Ω, four-wire, open-wire line with aluminium conductors of $\frac{3}{4}$-inch diameter has a power rating limited by a peak voltage of 2 kV/cm on the wires. This arrangement gives a factor of safety of about 3:1 for an HF carrier power of 500 kW (750 kW RMS average with 100% modulation and 2000 kW RMS peak power).

An advantage of all-open-wire transmission lines is that the location of faults and damage is easily seen. Open-wire lines are not standardized as regards outer

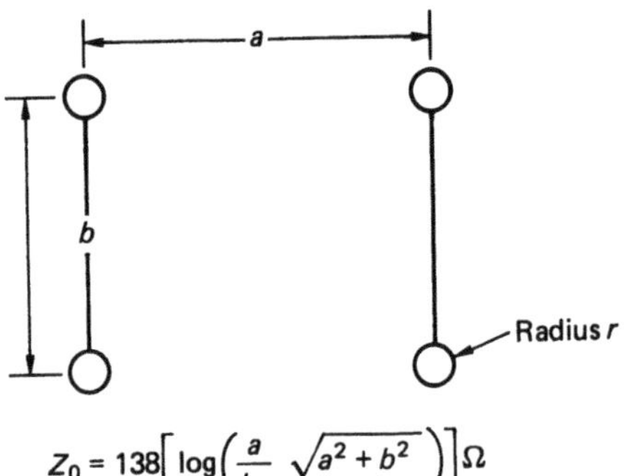

$$Z_0 = 138\left[\log\left(\frac{a}{br}\sqrt{a^2+b^2}\right)\right]\Omega$$

Figure 11.6 Four-wire, open-wire, balanced transmission line

Table 11.8

Type	*Band*	Z_0 *(Ω)*	*Carrier power (kW)*
Coaxial open-wire	LF/MF	230	100/200
Coaxial open-wire	LF/MF	120	2000
Balanced open-wire	HF	300	100
Balanced open-wire	HF	300	250
Balanced open-wire	HF	300	500

diameter size as are SAS and foam dielectric coaxial lines but a number of manufacturers offer a range suitable for high-power operation. As an example, a typical range of open-wire lines is shown in Table 11.8. The carrier powers shown apply under the extreme conditions encountered in tropical and high-altitude environmental conditions.

11.3.5 Waveguides

Waveguides are used as feeders in the SHF (microwave) band (2–22 GHz). Originally, waveguides were usually of the rectangular type shown in Figure 11.4 which can only be manufactured in short, rigid lengths. Rectangular guides are expensive and not well suited to applications where long, straight runs are not possible and bends are required. Most waveguides are now of the elliptical type which can be manufactured in long (semi-flexible) lengths that can be transported on drums and easily erected in situations where a number of bends are required. Rectangular waveguides are now used, mainly, as short connecting pieces, twists and sharp bends, and terminations are available that convert from the elliptical type. Rectangular waveguides are also sometimes used as feeders at UHF in situations where a very low loss is required. Here it is useful to note that the attenuation of a waveguide is substantially less than that of a coaxial line of the same external dimensions. Further, at very high frequencies, where the maximum diameter of a coaxial line is limited by the cut-off frequency, the external dimensions of a waveguide are greater than those of a coaxial cable, giving a further reduction in attenuation and increase in power rating.

In some situations, rigid, circular waveguides are used at frequencies above about 7 GHz and these, although more expensive than the elliptical type, have two advantages:

1. Because of the axial symmetry of a circular waveguide it is possible to transmit two orthogonally polarized transmissions (in effect one VP and one HP transmission) simultaneously. The isolation between the two trans-

missions is better than 30 dB and it is possible to transmit two different programmes via the same waveguide.

2. The transmission mode in a circular waveguide is such that it requires a larger cross-section area than that occurring with the elliptical type, operating at the same frequency. As a result, the loss is substantially lower but, since it is rigid and bends are not possible, a combination of circular and elliptical waveguides is usually necessary.

There are many manufacturers of rectangular, elliptical and circular waveguides and their products differ as regards precise dimensions and performance characteristics. However, as a guide, Table 11.9 lists commonly used elliptical waveguides and their approximate loss and power rating.

Table 11.9

Band (GHz)	*Size (mm)*	*Attenuation (dB/100 m)*	*Power rating (kW)*
1.7–2.3	145 × 83	1.3	60
2.5–3.1	112 × 67	1.6	30
3.1–4.2	84 × 17	2.4	15
3.6–4.2	72 × 51	3.0	12
4.4–5.0	57 × 43	4.5	6
7.1–8.5	45 × 28	6.0	3.5
9.0–10.0	34 × 24	8.9	1.9
10.7–11.7	33 × 21	9.1	1.8
12.2–13.3	29 × 19	11.3	1.2
14.4–15.35	26 × 17	13.8	1.0

It is interesting to compare the loss and power rating of a waveguide with those of a coaxial line operated in the SHF band. A $1\frac{1}{8}$-inch SAS coaxial line is the largest size that can be operated at 2 GHz and has a power rating of 3 kW and a loss of about 3 dB/100 m. On the other hand, a 145 × 83 mm waveguide has a loss of 1.3 dB/100 m and a power rating of 60 kW.

Table 11.10 gives the attenuation of typical circular, rigid waveguides and it will be seen that the attenuation is substantially lower than that of elliptical waveguides operating in the same bands.

Table 11.10

Band (GHz)	*Diameter (mm)*	*Attenuation (dB/100 m)*
6.0–8.5	83	2.7
10.7–13.0	31	4.2
14.0–15.0	27	5.0

11.3.6 Feeder selection

The range of bands, transmitter powers, types of modulation and environmental conditions that can arise in the selection of the optimum transmission line for a given application is obviously very wide. However, the method employed for the selection can be illustrated by an example of television in the VHF band. The specified requirements are:

Area	Middle East
Vision transmitter power	10 kW
Frequency	180 MHz

The operating factors described in Section 11.2.4 have been calculated and are shown in Table 11.11.

Table 11.11

Operating factor		*Value of operating factor* — *Peak power*	*Mean power*
Modulation	TV	$w_p = 1.7$	$w_m = 0.7$
VSWR	1.1:1	$v_p = 1.1$	$v_m = 1.1$
Ambient temperature	40°C	$t_p = 1.0$	$t_m = 1.0$
Solar radiation	Tropical	$r_p = 1.0$	$r_m = 1.14$
Altitude	100 m above mean sea level	$a_p = 1.01$	$a_m = 1.0$
Excess pressure	0 bars	$p_p = 1.0$	$p_m = 1.0$
Buried coaxial	Yes	$b_p = 1.0$	$b_m = 1.05$

From Table 11.11 the required standard ratings of the transmission line are:

$$\text{Peak power rating} = 10 \times w_p v_p t_p r_p a_p p_p b_p = 18.9\,\text{kW}$$
$$\text{Mean power rating} = 10 \times w_m v_m t_m r_m a_m p_m b_m = 9.2\,\text{kW}$$

From Table 11.6, a $1\frac{1}{8}$-inch diameter SAS transmission line would be suitable for the specified application. For a typical line length of 150 m the loss would be 0.9 dB but, if this is considered excessive, a larger diameter could, of course, be used. In this example, the transmission line is assumed to operate at atmospheric pressure but would need to be fed with dry air to prevent the ingress of moisture.

Selection of the optimum waveguide for an application is simpler than the selection of a transmission line. This is because the size is set by the required operating frequency and the power rating is therefore also determined. As has been pointed out, the power ratings of waveguides are relatively high and unlikely to impose any limitation in practice but it is normally necessary to pressurize with dry air to prevent the ingress of moisture.

Appendix 1

Contract administration

The placing of a contract involves an agreement between two parties, where the first party buys goods or services supplied by the second party. It is possible to use a number of descriptions for the parties: the first party can be called the buyer, end-user, customer, etc. and the second party can be called the seller, contractor, manufacturer, etc. In this appendix the first party is always referred to as the broadcaster and the second party is referred to as the contractor or manufacturer as appropriate.

A1.1 Turnkey contracts

In many countries, the lack of 'in-house' planning facilities makes the placing of contracts on a 'turnkey' basis very attractive. A 'turnkey contract' is a contract placed on a single contractor where that contractor provides, for the broadcaster, all of the equipment and services required for a project as a package. There are, basically, two types of turnkey contract, which are known as 'total turnkey contracts' and 'electrical turnkey contracts', and the difference is that, in the latter, civil works are excluded:

A1.1.1 Total turnkey contracts

The broadcasters of many countries have limited manpower resources for the planning and construction of their technical facilities. Indeed, in many cases, projects requiring transmission-engineering planning expertise only take place at intervals of some years and it would be quite uneconomic to employ staff whose efforts would seldom be required. This state of affairs results, at least in part, from the long life of transmitting equipment (20–30 years in the case of high-power transmitters).

As an example of a total turnkey contract it is useful to consider the provision of a medium-power (say 20-kW transmitter power) MF transmitting station. The broadcaster will, of course, need to issue a specification inviting qualified contractors to tender by a given date (a dated tender). He may write the specification himself, or alternatively arrange for it to be written by a specialist consulting engineer, but, in either case, the broadcaster must ensure that all his requirements are accurately stated. For a total turnkey contract the specification must cover the following.

(a) Civil works

The civil works include all buildings, outhouses, antenna tuning enclosures, workshops, roads, security fencing and antenna foundations. In principle, the civil works should also include water supply and main sewerage but often (e.g. in the case of a state-run broadcasting authority) the broadcaster may find it more convenient to be responsible for this provision.

(b) Electrical works

Electrical works include domestic (building) electrical power, air-conditioning, and the main electrical power supply for the transmission equipment, including standby power supply. The main electrical power supply will also include the power distribution and isolation switchboards. The broadcaster may, however, find it convenient to be responsible for the provision of incoming, high-voltage mains power from a national grid or local power station together with step-down transformer equipment for low-voltage supplies. Automatic voltage regulation equipment, however, is most conveniently provided by the turnkey contractor.

(c) Capital equipment

In the case of a 20-kW MF transmitting station this would include:

1. Transmitters and programme input equipment (PIE)
2. Studio to transmitting station programme links with order wires (command circuits)
3. Transmitter to antenna feeder
4. Antenna (mast radiator) and antenna matching unit
5. Test equipment, maintenance tools and workshop equipment
6. Standby 'announcer' studio facilities for use in case of link failure
7. Control room including remote control and monitoring facilities

The contractor would also be responsible for the installation and commissioning of all the above facilities to the satisfaction of the broadcaster. The contractor, therefore, is responsible for 'putting the station on the air' within an agreed and specified time-scale. As mentioned in Chapter 1, it is in the interests of the broadcaster to appoint his own 'project manager' to co-ordinate and monitor progress by the turnkey contractor.

A1.1.2 Electrical/electronic turnkey contracts

Quite often, especially when national broadcasting organizations are planning a project, it is a legal requirement that civil works must be carried out by the government Ministry of Works. In this case, therefore, two separate contracts are placed by the broadcaster:

1. A contract covering all civil works is placed on the Ministry of Works or on local private building companies via the Ministry of Works. This contract will cover buildings, roads, water, sewerage and mast foundations.
2. A contract covering all electrical and electronic work on the 'equipment/system contractor'.

In this 'split' contracting situation, the broadcaster has a more difficult role than he has in the case of a total turnkey contract, as he has to monitor the performance of two separate contractors. In particular, he must ensure that buildings, and other civil works facilities, are in a form suitable for the installation of the electrical and electronic equipment.

More often than not, the turnkey contract for equipment will be placed first, mainly because the building will have to be designed with a knowledge of the transmitting equipment in mind. It should be noted that, although all manufacturers offer similar transmitter powers, there are important differences in cooling requirements, electrical inputs, dimensions and mountings. Thus, once the turnkey contract for equipment is placed, the building design can commence and must be carefully monitored by the broadcaster. All too often, buildings have been erected with loading bays at the wrong levels, doors of the wrong width and ducting in the wrong place, necessitating expensive and time-delaying modifications. There is no doubt that splitting the work between a building contractor and an equipment contractor is less expensive than a total turnkey contract but it does require considerably more effort on the part of the broadcaster. The broadcaster must ensure that all the work is effectively co-ordinated and that the interfaces between the building and the technical equipment are correct.

It is even more economic for the broadcaster to split his purchases into even smaller blocks and to place separate contracts for transmitters, PIE, studio equipment, masts and antennas. In this case, however, the broadcaster will have a much more complicated task in co-ordinating all the separate contractors and their individual responsibilities. Such a course cannot, therefore, be recommended unless the broadcaster has available substantial staffing and engineering-planning facilities. It must be fully understood by the broadcaster that the more the purchases are split into separate parts, the greater is the degree of control, monitoring and co-ordination required from the broadcaster. In addition, and perhaps even more important, the greater is the responsibility of the broadcaster for the technical design of the complete system. With a total turnkey contract, the broadcaster has only one supplier and, hence, only one set of acceptance tests to carry out and, if there are any problems concerning equipment performance, he has only one contractor to deal with. This is obviously not the case if several, smaller, equipment contracts are placed.

A1.2 Contract terms and conditions

Most national broadcasting authorities have their own sets of contract conditions which are reviewed from time to time. On the other hand, system contractors and equipment manufacturers have their own sets of contract conditions. As it is most unlikely that these sets of conditions will be the same and as the two parties will have different sets of priorities, it is therefore important that, in the placing of a contract, both parties fully agree, and understand, the scope of their individual responsibilities. Agreed contract conditions can come about in two ways:

1. A set of conditions supplied by the broadcaster to which the contractor responds and proposes modifications he regards as necessary.
2. A set of conditions supplied by the contractor to which the broadcaster responds and proposes modifications within agreed government guidelines.

Whichever of these approaches is chosen there will normally be negotiations between the two parties to agree the overall wording of the final contract and its clauses. There are various standard points that must be checked by the broadcaster before he agrees that they are written into the contract:

1. *Definitions of the contract*. These are the definitions of various terms such as 'equipment', 'site location', 'engineer', etc.
2. *Testing and inspection*. Guidelines should be defined for equipment testing and inspection, the acceptance of equipment and acceptance protocols.
3. *Delivery*. Delivery dates should be shown, preferably in the form of a bar-chart. This may form the basis of a liquidated damages claim where the contractor is penalized for the late delivery of equipment. The broadcaster must ensure that he provides the correct information for shipment, for acceptance of equipment on site, and for the correct import documentation for the supplier.
4. *Equipment warranties (guarantees)*. Reputable suppliers normally provide guarantees for their equipment which, typically, is for a period of 12 months from ex-factory delivery or shipment as stated. It is possible to arrange for the extension of warranty but this must be negotiated on an item-by-item basis. The warranty normally applies to parts only and it is the responsibility of the broadcaster to notify the manufacturer as soon as a defective part is identified so that the manufacturer can then ship a replacement part. It may happen that a large item (e.g. a power transformer) becomes faulty and must be shipped back to the works for repair. It is very desirable, therefore, that the broadcaster maintains a stock of spare parts on which he can draw to avoid prolonged shut-downs.
5. *Liabilities on behalf of the supplier*. The supplier may, as part of his own terms and conditions of contract, arrange insurance for loss or damage to the equipment during shipping and installation. The scope of the insurance will, of course, depend on the extent of the contract (i.e. whether it is for

'supply only' or 'supply with installation'), and must be checked and understood by the broadcaster.

6. *Force majeure.* A *force majeure* clause is standard and is insisted on by the contractor in most supply contracts. This allows relief for the contractor if delays in delivery occur that are due to causes outside his control and the control of the broadcaster. Such causes will include war, flood, strikes and political unrest.
7. *Disputes.* Should there be a major dispute between the broadcaster and contractor concerning a contract, then, when all other avenues have been exhausted, it will be referred to arbitration. Such disputes between the contractor and broadcaster are referred, normally, to an arbitrator in a neutral area such as Switzerland or France.
8. *Equipment lists.* Schedules of equipment, clearly defined, must form part of any contract.
9. *Specification of equipment.* Specifications of equipment to be supplied should form part of the contract and will be used as a basis for acceptance tests.

In general, the purpose of contract conditions is to provide guidance for the broadcaster and contractor. The broadcaster wants to end up with a usable system and the contractor wants to provide a system acceptable to the broadcaster. Thus both parties want to achieve the same end and, with goodwill between the two parties, the execution of a contract should go smoothly.

A1.3 Fixed-price contracts and contract price adjustment

Most contracts are placed on a fixed-price basis where the contractor agrees to provide and install equipment for a fixed price with the broadcaster. Any adjustments to the price (e.g. for the provision of extra facilities) are negotiated separately and added as an amendment (or 'variation') to the contract.

There are some contracts (but these are becoming rarer) where a contract price adjustment clause is agreed that allows for an increase in prices. Contract price adjustment is normally relevant only in the case of long-term contracts, operating over a period of years, where the contractor cannot predict inflationary trends. In this case, separate formulae for equipment and services are agreed by both parties so that upward price adjustment can take place to compensate for inflation.

A1.4 Terms of payment and letters of credit

Most contractors have a set of standard payment terms that are included in their quotations and, eventually, included in the contract.

In most cases, 'down-payments' are required by contractors and the amount will depend on the size and time-scale of the contract. Down-payments usually vary from 10% to 50% but can involve a full payment of 100%. The reason for a high level of down-payment is usually so that the contractor can himself purchase equipment or parts on sub-contract and, in such cases, it is obviously in his interest to have available substantial working capital. Down-payments are normally paid by bank transfer, certified cheque or cash at, or soon after, contract signature.

The balance of payment will, in most cases, be provided on a 'letter of credit'. Contractors hesitate (with good reason) to deliver goods before receiving payment; broadcasters (with equally good reason) are reluctant to part with their money without having control of the goods. However, the matching of payments with deliveries is rarely possible, so that a compromise is normally agreed involving documented and constructive delivery (the handing over of documents transferring title to, or control of, the goods). Credit worthiness thus becomes important and bankers are required to intervene by giving their conditional undertaking to the contractor to pay against presentation of documents and of compliance with conditions stipulated by the broadcaster. To operate this arrangement there is clearly a need for documentary credits.

Documentary credits normally take the form of a 'letter of credit', supplied by the central bank of the country concerned and confirmed by a first-class overseas banking group. The 'confirmation' is required by the contractor, especially as he has little or no control over, say, foreign exchange releases by the country concerned. Letters of credit come in standard forms and allow the contractor to collect due payment against documented deliveries through bills of lading. Letters of credit are, in most cases, irrevocable; in other words, once in place, they cannot be rescinded, cancelled or amended.

The broadcaster will normally assume the ownership of goods once they are supplied (i.e. when documentary evidence such as a bill of lading is supplied). The contractor is paid at the time that he provides a bill of lading to the bank holding the letter of credit and, in this way, both the broadcaster and the contractor are satisfied. A standard form for a letter of credit is given at the end of this appendix.

A1.5 Delivery terminology

Today, with the development of new methods of transport and more sophisticated methods of documentation, there is a growing need for precise shipping and delivery definitions. It is important to know when to use a given term in a contract document and the consequences of the use of that term must be fully understood. The International Chamber of Commerce provides a comprehensive guide to the meaning of trading terms in their INCOTERMS publication. These definitions are widely accepted internationally and are available in many languages. It is impossible to list all the terms within the compass of this appendix but some of the most important are:

1. *Ex-works*. 'Ex-works' means that the costs given by the manufacturer assume that the broadcaster will collect the goods from his (the manufacturer's) premises. Thus the manufacturer will not be responsible for loading of goods, provision of vehicles and loss and damage during transit.
2. *FOB (free on board)*. FOB means that the costs given by the manufacturer include placing the goods on board a ship at a port of shipment named in the contract. Responsibility for loss or damage to the goods is transferred from the manufacturer to the broadcaster when the goods pass over the ship's rail.
3. *C&F (cost and freight)*. C&F means that the costs given by the manufacturer include the freight charges necessary to bring the goods to a named destination. Typically the destination could be the site at which installation is to take place. The risk of loss or damage, as well as any cost increases, is transferred from the manufacturer to the broadcaster when the goods pass over the ship's rail. From this point onwards, therefore, responsibility for insurance rests with the broadcaster.
4. *CIF (cost, insurance and freight)*. CIF is basically the same as C&F but with the addition that the manufacturer has to provide marine insurance against loss or damage to the goods. The manufacturer contracts with the insurer and pays the insurance premium. Ownership of the goods is transferred to the broadcaster when the goods pass over the ship's rail.

The above terms are those most commonly used in international shipment by ship or air and it is important that the broadcaster understands their precise meaning.

Adresse télégraphique
Cable address

No. de télex
Telex number

Lieu d'émission et date
Place and date of issue

CREDIT DOCUMENTAIRE - DOCUMENTARY CREDIT
IRREVOCABLE

Numéro du crédit / Credit number
de la banque émettrice / of issuing bank — *de la banque notificatrice* / of advising bank

Banque notificatrice / Advising bank

Donneur d'ordre / Applicant

Bénéficiaire / Beneficiary

Montant / Amount

Validité / Expiry
Date / Date
à / in — *pour négociation* / for negotiation

M / Dear Sir(s),

***Nous émettons en votre faveur ce crédit documentaire* / We hereby issue in your favour this documentary credit**

qui est utilisable par négociation de votre traite à
which is available by negotiation of your draft at

tirée sur
drawn on

et portant la mention: "Tirée en vertu du crédit documentaire No. — de (nom de la banque émettrice)"
bearing the clause: "Drawn under documentary credit No. — of (name of issuing bank)"

accompagnée des documents suivants:
accompanied by the following documents:

concernant
covering

Expédition/Embarquement de Despatch /Shipment from *à* to	*Expéditions partielles* / *Partial shipments*	*Transbordements* / Transhipments

Conditions spéciales:
Special conditions:

Nous garantissons aux tireurs et / ou porteurs de bonne foi que les traites émises et négociées en conformité avec les termes de ce crédit seront dûment honorées à présentation et que les traites acceptées conformément aux termes de ce crédit seront dûment honorées à leur échéance.

Le montant de chaque traite doit être inscrit au verso de ce crédit par la banque négociatrice.

We hereby engage with drawers and/or bona fide holders that drafts drawn and negotiated in conformity with the terms of this credit will be duly honoured on presentation and that drafts accepted within the terms of this credit will be duly honoured at maturity.

The amount of each draft must be endorsed on the reverse of this credit by the negotiating bank.

***Vos dévoués* / *Yours faithfully*,**

Nom et signature de la banque émetrice.
Name and signature of the issuing bank.

Indications de la banque notificatrice / Advising bank's notification

Lieu, date, nom et signature de la banque notificatrice.
Place, date, name and signature of the advising bank.

Sauf stipulations particulières expressément définies, ce crédit documentaire est soumis aux "Règles et Usances uniformes relatives aux Crédits Documentaires" (Révision 1962) Chambre de Commerce Internationale (Brochure No. 222).

Except so far as otherwise expressly stated, this documentary credit is subject to the "Uniform Customs and Practice for Documentary Credits" (1962 Revision) International Chamber of Commerce (Brochure No. 222).

Appendix 2

Definitions and formulae

A2.1 Logarithms and field strength

$Log_{10}(x)$ is shown as $\log(x)$ and natural logarithms, $\log_e(x)$, are shown as $\ln(x)$.
Basically, field strength is measured in volts/metre (V/m). However, at typical distances from a transmitting station, the value is normally well below 1 V/m and, for this reason, field strengths are usually quoted in either millivolts/metre (mV/m) or microvolts/metre (μV/m). For example, at the limit of the service area of an MF station the field strength is around 2 mV/m and the minimum usable field strength for a monophonic VHF/FM service is around 50 μV/m.
Field strengths of practical interest range from a few microvolts per metre to a few volts per metre (a range of a million to one or more). To provide a compact scale, covering this wide range, it is convenient to quote field strengths in decibels above one μV/m (denoted by dBμ). A field strength in dBμ is defined as $20\log(e)$, where e is the field strength in μV/m. For example, 10 μV/m becomes 20 dBμ and 500 mV/m becomes 114 dBμ so that, in this case, a voltage range of 50 000 : 1 is reduced to $114-20=94$ dBμ.

A2.2 Field strength and receiver input signal strengths

In principle, though not in practice at the longer wavelengths, the field strength of a transmission at any frequency at any point in space may be determined by varying the orientation of a half-wave dipole (with its fundamental resonance at that frequency) at that point and determining the maximum (RMS) open-circuit voltage, e, delivered by it. Conversely, if the field strength, E, is known, e can be determined. The relationship is defined by the formula:

$$e=E\times\frac{\lambda}{\pi}$$

If a receiver having the same input impedance, R, as the dipole were connected to it via a lossless feeder of similar impedance, the signal developed across the receiver input terminals, called the terminated signal, would be:

$$\frac{e}{2}$$

and the power delivered:

$$\frac{e^2}{4R}\,\mathrm{W}$$

Often, especially in the SHF band, received input signals are quoted in dBW, i.e. decibels relative to 1 W.

In an actual installation the signal delivered to the receiver would be multiplied by the gain of the antenna, relative to a half-wave dipole, less the loss in the connecting feeder.

A2.3 Gain and directivity

The power required to provide a given field strength at a given distance from an antenna depends on an antenna parameter known as gain, which is denoted by *G*. The gain of an antenna is defined as:

$$G = 10 \log (W_s/W_a) + 20 \log (E_a/E_s) \text{ dB}$$

where E_a is the field strength (on a voltage scale) produced by the antenna system, E_s is the field strength produced by a standard antenna, W_s is the power fed to the standard antenna and antenna system, and W_a is the power actually radiated by the antenna system.

For practical reasons, three different 'standard' antennas are used for expressing antenna gains in the different broadcast frequency bands:

1. In the LF and MF bands the standard antenna is a short monopole (a vertical radiator with a height small compared with a wavelength, i.e. less than 0.1 wavelengths) and gains are denoted by 'dBM'. For example, an MF mast radiator with a height of 0.55 wavelengths has a gain of 2 dBM.
2. In the HF and microwave bands the standard antenna is an 'isotropic source' (an imaginary antenna which radiates equal energy in all directions) and gains are denoted by 'dBi'. For example, an HRS 4/4/h HF curtain antenna has a gain around 21 dBi and a 3-m HF dish has a gain of 43 dBi at 6 GHz.
3. In the VHF and UHF bands the standard antenna is a half-wave dipole and gains are denoted by 'dBD'.

The standard antennas defined above are not always used as the standard antennas in the bands shown (e.g. the gains of MF antennas are sometimes quoted in dBi). In these cases, the following conversion factors can be used:

Gain in dBi = Gain in dBM plus 4.8
Gain in dBi = Gain in dBD plus 2.2

Gain in dBM = Gain in dBi minus 4.8
Gain in dBM = Gain in dBD minus 2.6

Gain in dBD = Gain in dBi minus 2.2
Gain in dBD = Gain in dBM plus 2.6

The gain of an antenna, in the absence of any losses, is referred to as the 'directive gain' or 'intrinsic gain', *G*. It is important to note that when the term 'gain' is applied to an antenna system it includes the losses in the antenna system. For example, if the intrinsic gain of a lossless antenna is 6 dBi and it is connected to a feeder with a 3-dB loss, the directive or intrinsic gain will be 6 dBi and the net gain 3 dBi. The net gain of an antenna system determines the field strength that is produced when a transmitter of given power is connected to the antenna system and is, therefore, particularly relevant to transmitting antennas. In the HF band, where external noise usually determines the signal-to-noise ratio (SNR) of a receiving system, the intrinsic gain of the antenna rather than the net gain of the system is relevant.

A2.4 Radiation patterns

A radiation pattern of an antenna is a plot of the radiated field of an antenna in cartesian or polar co-ordinates as a function of either azimuth (bearing) angle (the horizontal radiation pattern or 'HRP') or elevation angle (the vertical radiation pattern or 'VRP'). The radiated field can be plotted in terms of field strength or, alternatively, in terms of gain (or directive gain). When field strength is used, it is usually normalized and made equal to unity in the direction of maximum radiation. As an example, the HRP of a single quarter-wave MF monopole is a circle of unit radius (i.e. it is omnidirectional) and the VRP decreases from a maximum of unity at zero elevation to zero at 90° elevation. However, when a multi-element antenna is used to increase gain, both the HRP and the VRP will vary with angle and be a maximum at the required azimuth and elevation angles. It is important to realize that gain (or directive gain) is the result of a redistribution of the energy supplied by the transmitter. Gain (or directive gain) is the result of increasing the energy radiated on the required bearing and at a required elevation angle by reducing the energy radiated in other directions and at other angles.

A2.5 Impedance, VSWR and reflection coefficient

The input impedance of an antenna in general comprises a resistive term in series with a reactive term and the former accounts for the power radiated by the antenna (i.e. the power supplied to the antenna by the transmitter via a feeder). A transmitter is designed to deliver power into a specified terminating resistance (normally 50 Ω) and the impedance of the antenna (see through the feeder) must match this resistance within fairly close limits. At HF and higher frequencies the radiating elements of the antenna are made to be resonant and the impedance presented to the feeder is inherently resistive. At MF and lower frequencies this is not feasible in the antenna itself and transforming and reactance cancelling networks are necessary to achieve a reasonable match to the feeder. If a reasonably good match is not achieved, the transmitter may trip due to reflected power, the feeder may be subjected to unacceptably high voltages and antenna fault detection systems may not operate reliably.

In practice, the characteristic impedance of the feeder is chosen to be equal to the required transmitter terminating resistance and every attempt is then made to achieve an antenna impedance as close as possible to this value. The extent to which the antenna impedance may deviate from the required resistive value (normally 300 Ω balanced or 50 Ω unbalanced) is defined in terms of the VSWR (voltage-standing-wave-ratio) on the feeder connecting the antenna to the transmitter. When the antenna impedance is resistive and equals the characteristic impedance of the feeder, the energy from the transmitter (the forward wave) is all radiated by the antenna and the VSWR is equal to unity. However, when the antenna impedance differs from the required resistive value, energy is reflected back towards the transmitter (the reflected wave). The interaction between the forward and reflected waves sets up a voltage pattern on the feeder which varies between maximum and minimum values. The VSWR (which is a measure of the accuracy of matching of the antenna) is the ratio between the maximum and minimum values of voltage.

Useful formulae concerned with impedance and VSWR are:

$$\mathrm{VSWR} = (1+p)/(1-p)$$

where p is the reflection coefficient, which is the ratio of the amplitudes of the reflected and forward waves; and

$$p = (Z - Z_0)/(Z + Z_0)$$

where Z is the impedance of the antenna, and Z_0 is the characteristic impedance of the feeder (which is normally resistive and equal to the required transmitter terminating resistance). As examples, the maximum permissible VSWR for an HF broadcast antenna system is, usually, 2: 1 at the carrier frequency, and for an MF antenna it is around 1.05: 1 at carrier frequency, rising to 1.8: 1 at the extreme sideband frequencies.

A2.6 Bandwidth and beamwidth

The bandwidth of a circuit (a filter, amplifier, etc.) is the frequency range over which the response is substantially constant and falls by no more than a specified amount. Bandwidths are, therefore, referred to by the specified fall in response, e.g. a '3-dB bandwidth' or a '6-dB bandwidth'.

The 'bandwidth' of an antenna is the frequency range over which an acceptable performance can be obtained and is therefore limited by changes in the radiation pattern and impedance. The bandwidth imposed by changes in radiation pattern is known as the 'pattern bandwidth' and that imposed by changes in impedance is known as the 'impedance bandwidth'. Normally, a wide impedance bandwidth is more difficult to achieve than a wide pattern bandwidth, and therefore sets the operating bandwidth of an antenna.

The bandwidths of antennas in the various broadcast bands are, typically:

1. MF – Sufficient for a single AM transmission (some simple MF antennas can carry two transmissions separated by 10–15% in frequency)
2. HF – Wideband curtain 2:1 and log-periodic 4:1 or more
3. VHF/UHF – Up to 20 MHz for multi-programme Band II FM transmissions and 88 MHz for four UHF TV transmissions

The beamwidth of an antenna is normally defined as the angular range over which the radiated field strength falls by no more than a specified number of dB. This is referred to as the '3-dB beamwidth', '6-dB beamwidth', etc. It is a term most frequently used to describe the performance of SHF link antennas, HF antennas and, to a lesser extent, the vertical radiation patterns of UHF antennas.

A2.7 Transmitter power

Transmitter power is usually quoted in watts (W) or kilowatts (kW) but the following definitions and notes are useful:

1. In propagation calculations it is sometimes convenient to define power in decibels above 1 kW (dB/kW), given as:

$$\text{dB/kW} = 10 \log (W)$$

   where kW is the power in kilowatts.
2. For DSB (double-sideband) broadcast transmissions the stated transmitter power is the carrier power. For 100% modulation the maximum average power is 1.5 times the carrier power and the peak power (which occurs under 100% modulation conditions) is four times the carrier power.
3. For SSB (single-sideband) transmissions the stated transmitter power is the peak power under 100% modulation conditions and is referred to as PEP (peak effective power).

It should be noted that an SSB transmission with $W/2$ watts PEP will give the same received SNR as a DSB transmission with W watts carrier power. Under 100% modulation conditions the PEP of the SSB transmission will be $W/2$ watts and the PEP of the DSB transmission will be $4W$.

A2.8 Effective radiated power

So far as the received field strength is concerned, transmitter power and transmitting antenna gain are interchangeable, except at locations close to transmitting sites. For example, an MF transmitter power of 50 kW and an antenna gain of 3 dBM (a power ratio of × 2) will provide the same field strength as a transmitter power of 100 kW and an antenna gain of 0 dBM (a power ratio of × 1). The product of transmitter power in kilowatts and antenna gain (expressed as a power ratio) is called the ERP. The ERP determines the field strength and, for example, in both cases cited above, is 100 kW. The ERP can also be stated in decibel form with the transmitter power given in dB/kW and the antenna gain in dB. In the cases cited above the ERP is, therefore, 20 dB/kW, which derives from either 10 log (50) + 3 or 10 log (100) + 0. Note that in this case the ERP is termed the EMRP to indicate that the standard antenna is a short monopole and that the 'M' would be replaced by 'i' or 'D' if the antenna gain were given in dBi or dBD. For example, the ERP in the case of satellite transmissions uses the isotropic source as reference and is termed the EIRP.

Unless otherwise qualified, ERP refers to the ERP in the direction of maximum radiation.

A2.9 Antenna distribution feeder systems

Directional transmitting antennas use a number of radiating elements (usually dipoles or monopoles) driven by currents with the amplitudes and phases necessary to give the required radiation pattern. The distribution feeder system (i.e. the arrangement that distributes the transmitter output power to the radiating elements) will fall under one of four headings:

1. *Parasitic* in which one element is driven and the other elements receive energy by mutual coupling as illustrated in Figure A2.1. This arrangement is often used for directional MF antennas.

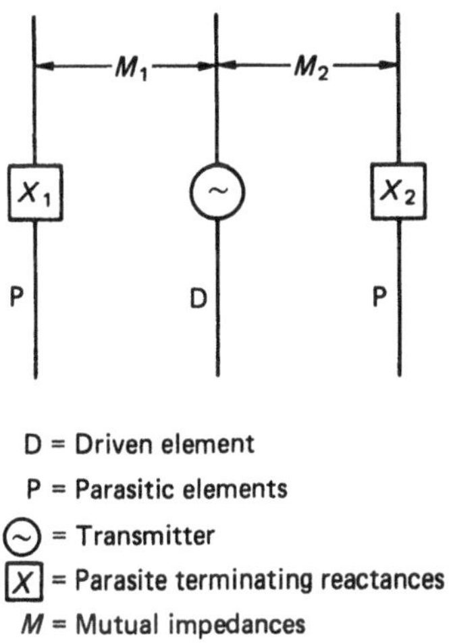

Figure A2.1 Parasitic antenna element feeds

2. *End-fed* in which the radiating elements are spaced along a transmission line with a length that provides the required current phase, as illustrated in Figure A2.2. This arrangement only operates over a narrow frequency band as the line-spacings will only be correct at one frequency.

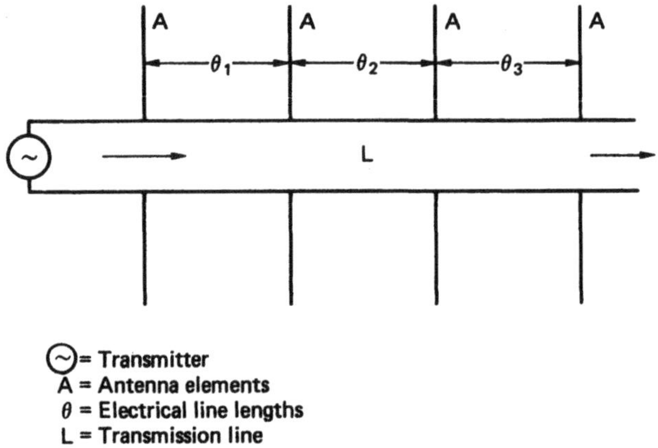

Figure A2.2 End-fed antenna elements

3. *Multiple-feed* in which the radiating elements are fed from a common point via feeders whose lengths and characteristic impedances give the required currents and phases as shown in Figure A2.3. The bandwidth of this arrangement is wider than that of the end-fed system and it is often used in low-gain VHF/FM and VHF/UHF television antennas.

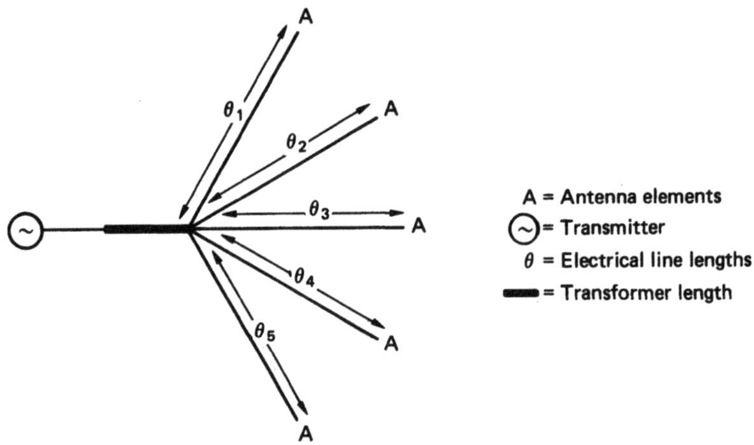

Figure A2.3 Multiple-feed antenna elements

4. *Branch-fed* which is illustrated in Figure A2.4 and is used where a large number of radiating elements have to be fed and it would not be electrically or mechanically possible to feed from a common point, or if unequal amplitude or variable phase feeds are required (e.g. to 'slew' the horizontal pattern of a wideband HF curtain antenna).

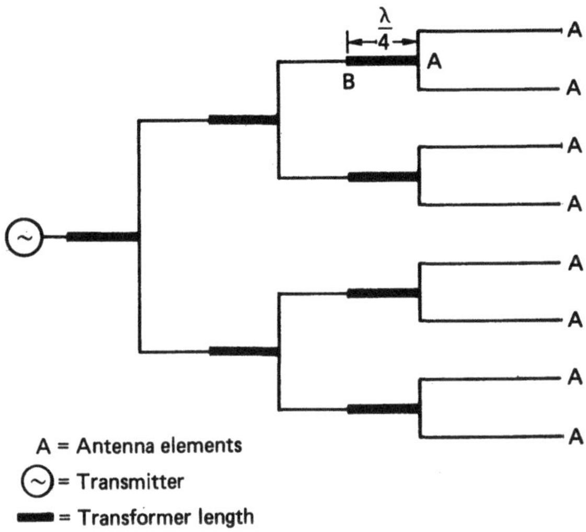

Figure A2.4 Equal amplitude co-phased branch feeder system

A2.10 Baluns and transformers

A balun is an arrangement for connecting an unbalanced transmission line to a balanced transmission line or vice versa. Figure A2.5(a) shows the effect of connecting a coaxial feeder (unbalanced line) to a dipole, a balanced device. The flow of charges travelling along the inside of the outer on emerging divides, some flowing in one arm of the dipole and some on the outside of the outer, in contrast to charges travelling on the inner, which all flow in the other arm of the dipole. The effect of this is to distort the radiation pattern of the dipole and change its radiation resistance, as currents on the outer surface of the outer conductor as well as those in the dipole arms radiate power. Figure A2.5(b) shows how a balun eliminates these undesirable effects: in effect charges travelling on the outside of the inner and inside of the outer of the coaxial transmission line now divide in a similar way, but there is no radiation of power from the balun because the currents in it are of equal amplitude, but flowing in opposite directions.

Transformers, which are usually a quarter-wavelength or a half-wavelength long (the latter usually being called double quarter-wave transformers) are used

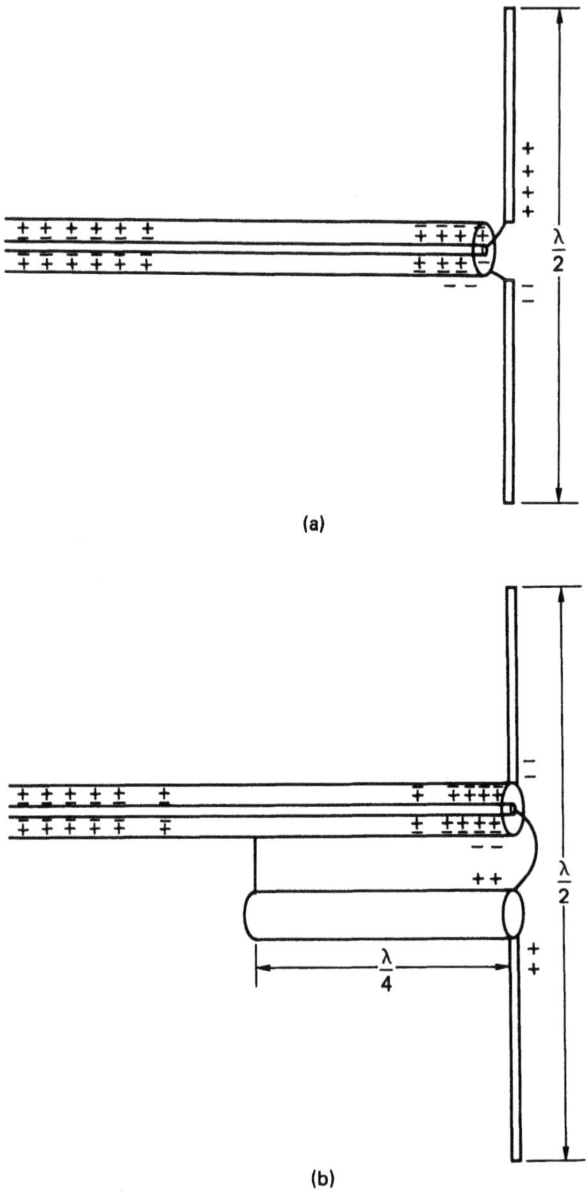

Figure A2.5 (a) Dipole and coaxial feeder without balun. (b) Dipole connected to feeder with balun

to transform the impedance at paralleling points so that feeders are terminated in their characteristic impedances. The formula for a single quarter-wave transformer of characteristic impedance Z_o is:

$$Z_1 Z_2 = Z_o^2$$

where Z_2 is the terminating impedance on the transformer, and Z_1 is the input impedance. As an example, refer to Figure A2.4, where all the feeders are assumed to have a characteristic impedance of 70 Ω. At point A the impedance will be 35 Ω and it is required to be 70 Ω at point B. Applying the formula $70 \times 35 = Z_o^2$:

$$Z_O = 49.5\,\Omega$$

In practice double or sometimes triple quarter-wave transformers are often used to achieve the desired bandwidth.

A2.11 Low power levels

It has already been pointed out that transmitter powers can be expressed on a logarithmic scale as dB/kW. For lower powers (e.g. those arising in satellite and receiver engineering) the same arrangement can be used with a reference of a watt or a milliwatt. For example, a power of 10 W can be expressed as 10 log (10) = 10 dBW or 10 log (10 000) = 40 dBm.

A2.12 Pre- and de-emphasis

Because in practice the amplitudes of the audio frequency (AF) components of programmes decrease with frequency it is feasible and usual in FM systems to amplify (pre-emphasize) the higher modulating frequencies before transmission and attenuate (de-emphasize) them, correspondingly, in the receiver demodulator, as this results in an improvement in the received SNR. A 'top-lift' network is included in the audio circuits of the transmitter and a 'top-cut' network in the receiver that, together, give an overall flat response. The receiver de-emphasis network has a transfer characteristic equivalent to that of a simple resistance–capacitance combination as shown in Figure A2.6. As will be seen, the transfer characteristic has an attenuation depending on the quantity RC which has the dimensions of time and is referred to as the 'time constant'. In VHF/FM

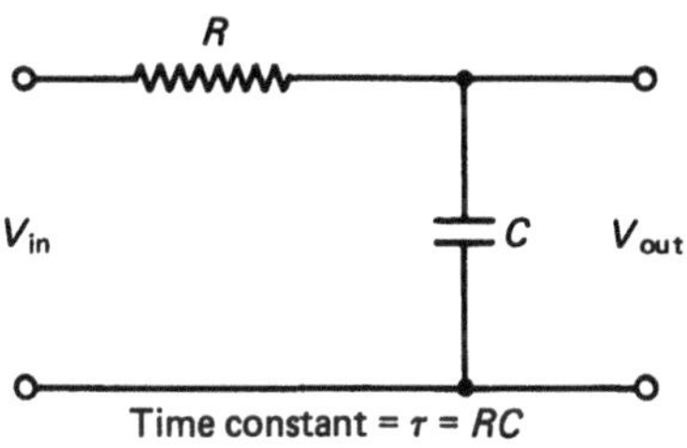

$$\frac{V_{out}}{V_{in}} = \frac{1}{(1+\omega^2\tau^2)^{\frac{1}{2}}}$$

$\omega = 2\pi \times$ frequency in Hz

Figure A2.6 De-emphasis network

transmission a time constant of 50 or 75 μs is usually used and results in a top-lift (and compensating top-cut) of either 10 or 14 dB.

A2.13 Radiation efficiency

Except in the LF and MF bands, the radiation efficiency is simply the power radiated by the antenna relative to the power input to the feeder–antenna system. Thus in the HF, VHF and UHF bands, a total loss of 3 dB (due to resistive losses in the antenna and feeder loss) corresponds to a radiation efficiency of 50%. However, the definition of radiation efficiency in the LF and MF bands is complicated by the need to distinguish between the losses in the neighbourhood of the antenna (which obviously affect radiation efficiency) and those at a distance from the antenna (which contribute to the path attenuation). There are a number of ways in which this problem can be solved and the one adopted in this book is to include losses up to one wavelength when determining radiation efficiency, with those beyond one wavelength contributing to path attenuation. The radiation efficiency depends on ground conductivity near the antenna and the number of radial wires used in the earth system. With a mast height of a quarter-wave or more, a typical earth system and typical earth conductivity and radiation efficiency is in the range 90–98% (a loss of 0.1–0.4 dB).

A2.14 Radio links

The free space (line-of-sight) path loss of a radio link or satellite transmission is defined as the ratio of input power to output power when isotropic transmitting and receiving antennas are used and is given by:

$$\text{Path loss} = 92.4 + 20\log(d) + 20\log(f)\ \text{dB}$$

where d is the path length in km, and f is the frequency in GHz. For example, a 12-GHz satellite transmission over a path length of 35 000 km will have a path loss of:

$$92.4 + 20\log(35\,000) + 20\log(12) = 205\ \text{dB}$$

The ratio of output power to input power with transmitting and receiving antennas with gains G_t and G_r dBi respectively is given by:

$$92.4 + 20\log(d) + 20\log(f) - G_t - G_r\ \text{dB}$$

The gain of a dish antenna is given approximately by:

$$G = 17.4 + 20\log(df)$$

where d is the dish diameter in m, and f is the frequency in GHz.

A2.15 Reception quality and noise

It might be thought that no matter how low the field strength of a particular transmission, sufficient antenna gain or receiver amplification would guarantee satisfactory reception, but this is not so. Practical considerations put an upper limit on the achievable gain of receiving antenna systems (and hence signals fed to receivers) and there will inevitably be spurious signals falling within the frequency bandwidth of the desired transmission. Spurious signals may be natural or man-made, e.g. lightning, galactic noise and arcing in electrical equipment, or other transmissions on the same frequency. Clearly the desired transmission must be greater than any spurious signals by a defined amount determined by the mode of transmission and quality of service to be provided. However, even if it were possible to eliminate these sources of interference there is a different type of interference, known as noise, generated in every type of receiver, resulting from the atomic nature of matter and the thermal agitation of electrons in resistances. The mean square value of this thermal noise voltage (e.m.f.) is:

$$E^2 = 4RkTB$$

where E is in volts, R is in ohms (input impedance of receiver), T is the absolute temperature, B is the bandwidth of the transmission channel in Hz, and k is Boltzmann's constant, equal to 1.38×10^{-23}. See also Section 10.3.2.

In any practical receiver the noise generated in the receiver always exceeds that given by the above formula by a factor which is defined as the noise factor of the receiver. This is sometimes quoted in dBs and is then normally referred to as the noise figure of the receiver. Thus the noise performance of receivers determines the minimum input signals to achieve the desired SNRs, assuming the quality of the received signal is not the limiting factor.

In practice, at LF, MF and HF, largely due to lack of directivity of receiving antennas, spurious signals and modes of propagation rather than receiver noise determine the quality of reception, but increasingly with VHF, UHF and SHF the noise performance of receivers is the controlling factor.

The relevant formulae for 625-line Standard I vision transmissions are:

$$\text{SNR (dB)} = F + G - 20 \log f + 23.97 - N$$

where F is the field strength (RMS amplitude of vision carrier during line-synchronizing pulses) in dB relative to 1 μV/m, G is the net gain of antenna feeder system in dB relative to a half-wave-dipole, N is the noise figure of receiver in dB, and f is the frequency in MHz; and

$$\text{SNR (dB)} = V - 8.118 - N$$

where V is the terminated signal (RMS amplitude of vision carrier during line-synchronizing pulses) in dB relative to 1 μV (see Section A2.2), and N is the noise figure of the receiver in dB.

For high-quality television and SHF receivers *N* would be typically 3.5 dB and 1–2 dB, worse if used with input filters, as is usual. The required SNRs for differing standards of a television service are quoted in Section 5.4.5.

A2.16 Electromagnetic waves, polarization and fields

The radiation of energy in the form of electromagnetic waves, which includes RF transmissions, was predicted by James Clark Maxwell in 1862, but the existence of such waves was not demonstrated until 1887 by Heinrich Hertz. Maxwell summarized the then known effects of electric currents and magnetism in a series of equations, and applied them to a closed volume of space. In so doing he derived another equation in which the term on the left-hand side corresponded to electrical energy being fed into the volume of space, whilst on the right-hand side there were four terms. Three of these terms he recognized as the production of heat and the storage of energy in electric and magnetic fields respectively. He then applied the principle of conservation of energy, reasoning that the fourth term, which was similar in form to that describing wave motion in elastic solids, represented power being radiated from the volume of space in the form of waves comprising electric and magnetic fields. Electric and magnetic fields are defined as regions of space where an introduced electric charge or magnetic pole, respectively, would experience a force. The magnitude of the force is a measure of the strength of the field at that point in the direction in which the charge or magnetic pole would move under the influence of the force in the direction of the field, or line of force, at that point.

Maxwell was further able to show that in an electromagnetic wave the electric and magnetic fields must exist together, be synchronized in amplitude and time, and be at right-angles to each other and to the direction of propagation of the wave.

By convention the polarization of an electromagnetic wave is defined by the direction of the electric field. Electromagnetic waves can be either plane polarized or elliptically polarized, of which circularly polarized is a special case.

Plane-polarized waves are described as vertically, horizontally or slant polarized, depending on the orientation of the electric field relative to the earth (a horizontal plane). A vertical dipole antenna in space produces vertically polarized waves on all bearings as illustrated on a particular bearing in Figure A2.7.

In a plane-polarized wave the electrical magnetic fields rise and fall sinusoidally as the wave propagates through points A in Figure A2.7. In a circularly polarized wave the electric and magnetic field strengths do not vary in amplitude but the directions of the fields rotate. From E to A, the plane of the electric field rotates from the plane xz to the plane yz and the plane of the magnetic field rotates from the plane yz to the plane xz. The fields can rotate in either direction and waves can thus be either left- or right-hand circularly polarized and it is thus important to clearly define the type of polarization

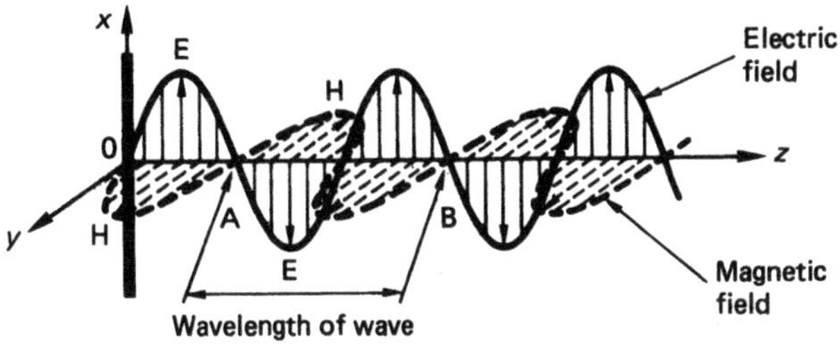

Figure A2.7 Graphical representation of a plane wave

requested or already in use. The usual convention is that, when looking at a transmitting antenna, and the electric vector rotates in a clockwise direction, the wave is defined as right-hand circular polarized. Elliptically polarized waves are intermediate between circularly and plane-polarized waves, in which the amplitude of the fields vary as they rotate.

A2.17 Radiation: public and occupational exposure levels

Radiation at frequencies used for broadcasting and SHF links is non-ionizing, i.e. it causes atoms and molecules to vibrate (thus generating heat) rather than breaking them apart, and there is little or no evidence to suggest that there are any risks to individuals apart from those arising from excessive general, local or surface heating of the body.

At frequencies between 100 kHz and 10 MHz the main hazard is the risk of localized burns resulting from contact with unearthed metallic objects, whilst above 10 MHz it is localized deep or general overheating of the body.

Experiments have demonstrated that behavioural changes in animals are not observed until the intensity of radiation (electromagnetic fields) to which they are exposed results in average specific energy absorption rates of 2–8 W/kg of body weight, and that those behavioural changes do not persist when exposure ceases. Furthermore, direct human experience from therapeutic use of electromagnetic fields indicates that localized levels of up to 20 W/kg for short periods can be accommodated by the thermoregulatory mechanisms of the human body.

On the basis of the above, an International Non-ionizing Radiation Committee has agreed upon a number of restrictions, some of which are that the average specific energy absorption rate in the body over any six minutes should not exceed 0.4 W/kg for occupational exposure, but for any 0.1 kg of tissue in the hands, wrists, feet and ankles it may be as high as 20 W/kg and for other parts of the body it may be up to 10 W/kg.

Figure A2.8 shows the field strengths corresponding to the 0.4 W/kg of body weight for occupational exposure. The minimum in the graph is due to greater absorption of energy at frequencies at which the human body is resonant.

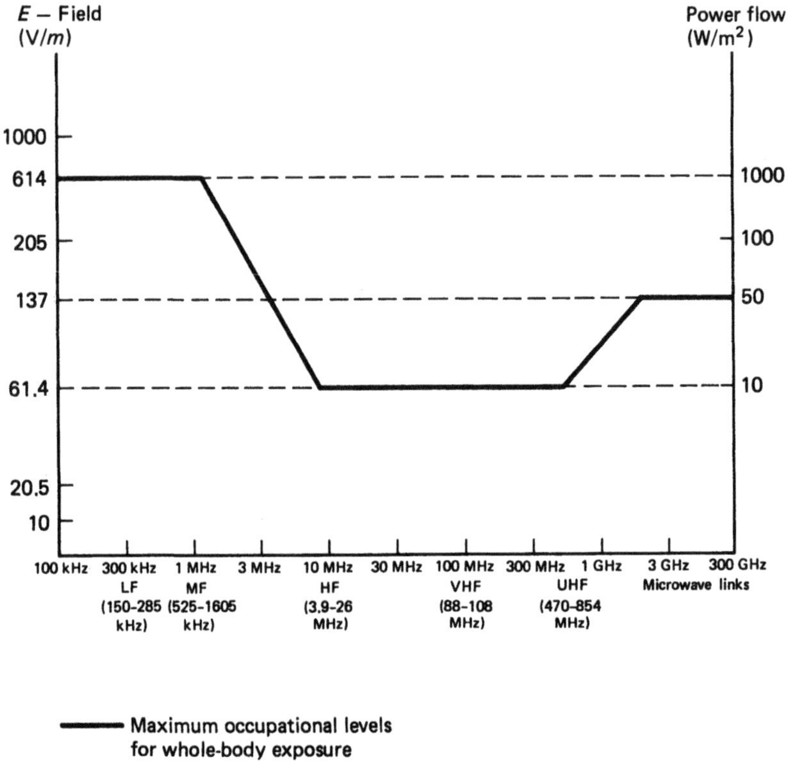

Figure A2.8 Exposure guidelines covering broadcast bands

The restrictions of the Committee thus include a factor of safety of between 5 and 20 on the results obtained from animal experiments for occupational exposure and a further factor of at least 5 is specified for public exposure, as a precaution against possible effects as yet unobserved.

Radiation hazard-measuring instruments are commercially available, but some experience is necessary in their use, as near-field effects can result in significant measuring errors with some types of instruments.

A2.18 Frequency bands

Table A2.1 shows the frequency ranges of the various bands, but it should be noted that only parts of these bands have been allocated for broadcasting and programme links. For completeness the wavelength ranges are also quoted.

Band	*Frequency range*	*Wavelength range (m)*
Low frequency LF	30–300 kHz	10^4–10^3
Medium frequency MF	0.3–3 MHz	1000–100
High frequency HF	3–30 MHz	100–10
Very high frequency VHF	30–300 MHz	10–1
Ultra high frequency UHF	0.3–3 GHz	1–0.1
Super high frequency SHF*	3–30 GHz	10^{-1}–10^{-2}

*Frequencies above 1 GHz are usually referred to as microwaves.

Frequency and wavelength are related by the formula:

$$C = n\lambda$$

where C (velocity of light) = 3×10^8 m/s, n is the frequency (Hz/s), and λ is the wavelength (m).

Index

www.ingramcontent.com/pod-product-compliance
Ingram Content Group UK Ltd.
Pitfield, Milton Keynes, MK11 3LW, UK
UKHW021505240926
475702UK00035B/444
9781032622484